George Washington Tryon

A Monograph of the Terrestrial Mollusca Inhabiting the United States

George Washington Tryon

A Monograph of the Terrestrial Mollusca Inhabiting the United States

ISBN/EAN: 9783337805647

Printed in Europe, USA, Canada, Australia, Japan

Cover: Foto ©berggeist007 / pixelio.de

More available books at **www.hansebooks.com**

For Terms See Fourth Page of Cover.

A MONOGRAPH

OF THE

TERRESTRIAL MOLLUSCA

INHABITING THE

UNITED STATES.

With Illustrations of all the Species.

BY GEORGE W. TRYON, JR.,

Editor of The American Journal of Conchology ; Member of the Academy of Natural Sciences of Philadelphia; Corresponding Member of the Boston Society of Natural History, the Lyceum of Natural History of New York, the California Academy of Natural Sciences, the Zoologischen-botanischen Gesellschaft in Wien, etc.

Part

PHILADELPHIA:
Published by the Author, 625 Market Street.

NEW YORK: LONDON:
Baillière Brothers No. 520 Broadway.   Trubner & Co., No. 60 Paternoster Row.

PARIS   MADRID:
J B Baillière et Fils, Rue Hautefeuille.   C. Bailly-Baillière, Calle del Principe.

BERLIN:
Asher & Co., No. 20 Unter d. Linden.

FRESH WATER SHELLS WANTED.

CONCHOLOGISTS or Dealers having FRESH WATER SHELLS from any part of the world, for sale or exchange, will please forward Catalogues of species and prices, to

CHARLES M. WHEATLEY,

Jan. 1865, (1y) Phœnixville, Pennsylvania.

GEORGE W. TRYON, Jr.,
625 MARKET STREET,
PHILADELPHIA, U. S.

EXCHANGES.

I offer to Conchologists and Collectors, to exchange any of my publications, including the "*American Journal of Conchology,*" and also my duplicate shells, for their memoirs on Conchological subjects or for specimens.

SHELLS WANTED.

I desire also to purchase species new to my collection, and request gentlemen having shells for sale to forward to me Catalogues with prices annexed.

SHELLS FOR SALE.

My stock of duplicates numbers nearly 3000 species. I will dispose of them in preference by exchange, or will sell them at very low prices. The land and fresh water shells of Europe and the United States are largely represented among these duplicates.

I can offer the following collections:

Land and Fresh Water Shells of Europe, 300 species		$ 50 00
do	200 species	20 00
do	100 "	10 00
Clausilia, 100 species, 250 specimens		15 00
" 50 " 100 "		5 00
American and Exotic Unionidæ, 100 species		25 00
" " " 50 "		10 00
" Fresh Water Univalves, 300 species		50 00
" " " " 200 "		25 00
" " " 100 "		10 00
Helix, 100 species		20 00
Marine Shells, 1000 species		250 00
" " 500 "		100 00
" " 250 "		35 00

Shells for sale also singly at low rates.

NAMING SPECIMENS.—I offer to NAME collections of Shells, for the privilege of retaining, (in exchange,) Shells that prove to be new to my Cabinet.

CONCHOLOGICAL WORKS FOR SALE.—See list on second and third pages of cover of American Journal of Conchology for October, 365.

DIRECTIONS FOR SHIPPING.—Gentlemen residing in Europe are requested to direct to me as above, with the addition:—

'In care of the Smithsonian Institution, Washington, D. C.,"

and forward their packages to either of the following booksellers:—G. Bossange & Co., Paris, Dr. Felix Flugel, Leipzig, Saxony. Wm. Wesley, 2 Queen's Head Passage, Paternoster Row, London.

American correspondents can forward heavy packages by express or rail. Light packages can be forwarded by mail.

AMERICAN JOURNAL OF CONCHOLOGY.
VOL. I.—1865.

Containing 400 pages, illustrated by 31 colored and plain plates; with papers by Conrad, Anthony, Bland, Wheatley, Binney, Newcomb, Haldeman, Stimpson, etc. Price in numbers, $12. For each number separately, $3.60. A few copies, beautifully bound in red morrocco, gilt, bevelled sides, red edges, at $15.

NOTICE.—To those who subscribe to Vol. II., the First Volume will be furnished at $10, or bound as above, at $12.

A MONOGRAPH
OF THE
TERRESTRIAL MOLLUSCA
INHABITING THE
UNITED STATES.

With Illustrations of all the Species.

BY GEORGE W. TRYON, JR.,

Editor of The American Journal of Conchology; Member of the Academy of Natural Sciences of Philadelphia; Corresponding Member of the Boston Society of Natural History, the Lyceum of Natural History of New York, the California Academy of Natural Sciences, the Zoologisch-botanischen Gesellschaft in Wien, etc.

PHILADELPHIA:
PUBLISHED BY THE AUTHOR, 625 MARKET STREET.

NEW YORK:
BAILLIERE BROTHERS No. 520 Broadway.

LONDON:
TRÜBNER & Co., No. 60 Paternoster Row.

PARIS:
J B BAILLIERE ET FILS, Rue Hautefeuille.

MADRID:
C. BAILLY-BAILLIERE, Calle del Principe.

BERLIN:
ASHER & Co., No. 20 Unter d. Linden.

MONOGRAPH

OF THE

TERRESTRIAL MOLLUSCA

OF THE

United States.

PREFACE.

The substance of the following pages was written by me some years ago for my own use, in identifying the species of North American Land Shells. Having found it an important aid in this connection, saving me much time, trouble and uncertainty, I have concluded to publish it, in the hope that it will prove alike useful to others pursuing the same study. To this end, the subject has been carefully reviewed, much of it re-written, and all the species published to this date included.

After a vain attempt to reconcile the prominent characters of the shells with the characters of the soft parts, or animals, as described by recent observers, and to divide the species according to the families and sub-families which they have instituted, I have concluded to introduce only such of these divisions as will give the genera and species as nearly as possible in their natural sequence, founded upon *Conchological* distinctions.

In pursuing this course (unavoidable, on account of the paucity of our knowledge of anatomical characters, and of the values of these characters in classification), we less regret the absence of a strictly scientific arrangement, because without it we can more readily attain the object of these pages than by separating, according to supposed important characters of the soft parts, species which, conchologically, greatly resemble one another, and again, on the same grounds, uniting incongruous forms.*

As nearly all the proposed genera of the American Helices appear to be well founded upon differences of animal and shell, I have adopted them, believing that they will facilitate rather than embarrass the investigator.

Finally, I may be permitted to add, these pages being written, not so much for the use of those who *are*, but rather for those who *desire to become*, Conchologists, it has seemed to me most proper that the descriptions of species should be as concise and as free from technical words as possible; that even characters of minor importance, and those, especially, which require microscopic observation, should be generally omitted, or only slightly alluded to; that the would-be naturalist

---

* Several recent writers have separated from the *Helicidæ* certain species, which otherwise are entirely undistinguishable from that family, on account of the animal possessing a mucous pore near its tail, like the snails of the genera *Arion* and *Parmacella*. They have accordingly united them either to *Arionidæ* or *Parmacellidæ*.

The late M. Moquin-Tandon, a most excellent French malacologist, appears to us to have correctly determined the presence of a developed spiral shell as of sufficient importance to justify the retention of these species within the family *Helicidæ*.

The possession of this mucous pore does not appear to characterize the animals of any particular groups of our species, for in those most nearly allied it is absent in one and present in another, and of a well characterized group of species, one only has been found to possess it. Very many of these animals have not yet been examined; so that, at present, it would serve no useful purpose to use this distinction in classification.

The differences in the jaw (buccal plate) and the armature of the tongue (lingual dentition) are employed by recent investigators for the combination of the genera of the Helices into sub-families, several of which appear to form good natural divisions, as *Helicellinæ* for the thin, glabrous species with unreflected lip. *Helicinæ*, on the contrary, includes a heterogenous collection of forms, from which I would suggest that the species of *Patula* should be separated, though I do not think they can possibly be correctly associated with *Vallonia* and *Planospira*, *Strobila* and *Helicodiscus*.

A close study of all that has been accomplished by American malacologists convinces me that —

1. If the sub-families proposed are properly characterized, *i. e.*, are natural groups, it is impossible, with our present limited knowledge, to properly distribute all the species among them, and —

2. For this reason, other sub-families must remain to be characterized; but —

3. If, when we obtain a knowledge of the characters of the soft parts of the at present *unarrangeable* groups of species, it does not show the existence of other sub-families, then, in all probability, *no* sub-families, as now defined, exist.

The able investigators, Messrs. Binney, Bland and Morse, who are now carefully and thoroughly studying the dentition of our species, will doubtless, in due time, arrive at results in the highest degree satisfactory to conchological students. Meanwhile, impressed by the unsatisfactory nature of our own edifice (equally with that erected by our predecessors), we trust that it will remain unquoted by future systematists.

should *first* be induced to *collect* and *name*, *afterwards* to arrange his collections upon the study of their habits and affinities; and that naturalists would become ten times more numerous were they not appalled at the outset by the *immensity* of the subject, which afterwards proves its greatest charm.

There are, besides, many persons whose leisure or tastes do not permit them ever to become, strictly speaking, naturalists, but who, nevertheless, if furnished with proper guides, would gladly devote a few hours occasionally to the pursuit of some branch of natural history, although they would never engage in it to such extent as to make it a *study;* and we feel that this large class of persons should be encouraged to do as much as possible for the benefit of science by collecting the material necessary for the researches of the educated naturalist.

※
※　※

Whenever good specimens could not be readily obtained for figuring, recourse has been had to previously published figures, and my acknowledgments are due to the works of Messrs. Binney, Bland, Lea and others, for the opportunity afforded by them for completing the illustrations. Very many of the shells are figured for the first time, as this is the first complete monography published since that of Dr. Binney.

The large number of new species described since that period have been classified in accordance with their affinities with those previously characterized.

At the close of the work will be found a list of American works on terrestrial species, and also a systematic list of the species and their principal synonyms. This has been deemed far preferable in the present instance (being more in accordance with the plan and purpose which I have endeavored to explain), to the usual method of prefixing the synonymy to the description of each species.

GEORGE W. TRYON, Jr.

625 Market Street,
Philadelphia, July 1st, 1866.

INTRODUCTION.

* * *

The number of described species of terrestrial mollusca known to inhabit within the limits of the United States is not far from 275 species, and many of these have a geographical distribution almost co-extensive with our territorial limits. There are, however, some peculiarly Southern and Californian groups of more restricted distribution, while there are not wanting, in our far Southern and South-western States, stragglers from the peculiar faunas of Mexico and Cuba; and even a few European species have become domesticated with us.

While *all* the species discovered and described in the United States to this date are included in the following pages, a few Mexican species are also added, generally as illustrations of allied forms inhabiting an adjacent country, many of which will doubtless be detected within our boundary when our little-known contiguous territories are more fully explored by the naturalist. A brief account of the geographical distribution of the species and of their habits is included under the description of each genus.

In the description of the species, for the sake of conciseness, many characters are omitted which are common to, and have already been used in the description of any of the groups, genera, or higher divisions in which the species is included, and, therefore, no specific description can be regarded as entirely complete in itself. It will, of course, be readily understood that adult shells of perfect growth are alone selected for these descriptions, and the same species, when immature, presents a very different aspect. *Mesodon abbolabris*, an animal of which the adult is furnished with a shell having a reflected and appressed lip and covered umbilicus, is, when young, furnished with a sharp-edged lip, and the umbilicus is open. The observer will soon, by a few comparisons, be enabled readily to separate the mature from the young shells, as the latter always present a peculiarly unfinished, fragile aspect.

When the ascertained geographical distribution of a species is extensive, we have generally indicated merely the States forming its boundaries. Its occurrence may be presumed within all the included region. More particular localities are only stated when the hitherto ascertained habitat of a species is extremely local.

Measurements are made in millimetres, 25 = 1 inch.

Key to the Families of Terrestrial Mollusca inhabiting the United States.

Order PULMONIFERA.

Section A.—ADELOPNEUMONA (Inoperculata).

Includes both terrestrial and fluviatile species.
Shell without operculum.

Sub-order GEOPHILA.

Terrestrial shells. Head furnished with *four* tentaculæ, either retractile by inversion or contractile, the superior pair with the eyes at their summits. Respiratory orifice closed by a valve.

* TENTACLES RETRACTILE BY INVERSION.

† *Shell external.*

‡ Shell thin, polished, translucent, lip sharp-edged.

§ *Shell oblong-oval or fusiform.*

Family OLEACINIDÆ.

The species of *Oleacinidæ* inhabit the vicinity of sea-shores of the Southern States from South Carolina to Texas; also the Pacific coast, but not extending so far northwards as California. They are all large in size, and are stragglers from the Mexican and West Indian faunas.

§§ *Shell obliquely oval, whorls very few, rapidly increasing in size, the last one constituting nearly the entire shell.*

Family SUCCINIDÆ.

Distribution universal.

§§§ *Shell discoidal, suborbicular, turbinate or trochiform.*

Family HELICELLIDÆ.

Inhabit throughout the United States, but mostly in the Northern States east of the Rocky Mountains.

‡‡ Shell thicker, striate, frequently banded with colors, lip thick-edged, or generally either with an interior thickening caused by a marginal deposit of callus, or else reflected outwards and appressly flattened.

§. *Shell discoidal, orbicular, turbinate or trochiform.*

Family HELICIDÆ.

istribution universal.

§§ *Shell oblong or oval.*

Family ORTHALICIDÆ.

Tropical, a few species inhabiting the Gulf States.

§§§ *Shell cylindrical or pupæform.*

Family PUPADÆ.

Distribution universal.

†† *Shell internal, rudimentary, contained beneath the mantle.*

§ *Shell small, flattened, not spiral. Respiratory orifice in the posterior right margin of the mantle.*

Family LIMACIDÆ.

Distributed throughout the States east of the Rocky Mountains, and one species in Oregon. Most of the species are European in origin, and only occur near the coast, especially in the vicinity of the large commercial cities.

§§ *Shell represented only by a few agglomerated calcareous granules. Respiratory orifice in the anterior of the right margin of the mantle.*

Family ARIONIDÆ.

A single European species is found in seaports of the Northern States.

††† *Shell entirely absent. Mantle covering the entire upper surface of the body of the animal.*

Family PHILOMYCENIDÆ.

There are two species known, which inhabit the Northern and Middle States.

** TENTACLES (AND HEAD) CONTRACTILE BENEATH THE MANTLE, WHICH COMPLETELY COVERS THE BODY OF THE ANIMAL.

§ *Mantle smooth, coriaceous.*

Family VERONICELLIDÆ.

Tropical. One species only occurs in Florida.

§§ *Mantle tuberculate.*

Family ONCHIDIIDÆ.

A single species is found in Lower California.

Sub-order LIMNOPHILA.

Amphibious and fluviatile species. Head furnished with *two* tentaculæ, with eyes at their *bases*.

A few of these shells are considered terrestrial, because they inhabit land within reach of the tides.

> § *Shell oval-oblong or conoidal, mouth dentate within. Not umbilicate.*

Family AURICULIDÆ.

The American are all *sea-shore* species, except one minute shell (*Carychium exiguum*) which extends far inland, upon river margins and in damp places.

Section B.—PHANEROPNEUMONA (Operculata).

Shell operculate. Head furnished with two tentaculæ, with eyes sessile at their bases. Respiratory orifice without a valve.

Sub-order ECTOPATHALMA.

Eyes lateral, at the external bases of the tentacles. Operculum spiral or concentric.

> § *Shell orbicular or turbinate, whorls convex, lip continuous, reflected, umbilicus open or indented.*

Family CYCLOPHORIDÆ.

A few species inhabit the Gulf States.

> §§ *Shell depressed, conical, whorls but little convex, lip not continuous, reflected, umbilicus covered by a heavy deposit of callus.*

Family HELICINIDÆ.

Tropical. Several species in the Gulf States, and one found as far north as Indiana. None in California.

Sub-order OPISOPTHALMA.

Eyes placed posterior to the bases of the tentacles. Operculum sub-spiral.

> § *Shell cylindrical, apex truncate.*

Family TRUNCATELLIDÆ.

Inhabiting sea-shores of Florida and California.

MONOGRAPH
OF THE
TERRESTRIAL MOLLUSCA
OF THE
United States.

Family OLEACINIDÆ, H. and A. Adams.

A single genus of this family inhabits the United States.

GLANDINA, Schumacher.

Shell oblong-oval, more or less elongated, corneous, shining; spire elevated; body whorl attenuated a little at base, imperforate; columella thin, arcuated, abruptly truncated at base; mouth long, rather narrow, lip sharp.

Animal, mouth capable of a proboscidiform protrusion, without a jaw.

The species of *Glandina* present few prominent distinctive characters, and are particularly liable to variation in size and form. As already mentioned, our species are southern in distribution, and occur only upon the sea-coast or in its near vicinity.

* *Transversely striate, without revolving striæ.*
† *Oval, cylindrical, spire moderate.*

1. **Glandina truncata,** Say.
Plate 1, figures 1, 2.

Ovate-fusiform, heavy, striate; spire rather elevated; apex obtuse, suture well marked, crenulate; whorls 6–7, moderately convex; body three-fourths the total length, sub-cylindrical; aperture ovate-lunate, narrow, half the total length, labium sharp, rounded below to meet the perpendicular, truncate columella. Fawn color tinted with rose, deeper within the aperture.

Length 65, diam. 25 millimetres.

South Carolina to Florida.

2. **Glandina parallela,** W. G. Binney.
Plate 1, figure 3.

Oval-cylindrical, very solid, with numerous delicate striæ; spire elevated, obtuse, suture moderate; whorls 6-7, convex; body *with straight parallel sides.* White.

Length 56, diam. 20 mill.

Louisiana.

---

3. **Glandina Texasiana,** Pfeiffer.
Plate 1, figure 4.

Oblong, elongated, narrow, striate, shining, pellucid; spire convexly conical, obtuse, suture pallid, minutely crenulate; whorls 6, slightly convex; body longer than spire, attenuated at base; mouth half the total length, narrow, labrum flatly rounded, columella perpendicular, or *generally arcuate.* Yellowish rose color.

Length 29, diam. 10·5 mill.

Texas.

Narrower, smaller, and proportionally more solid than No. 1.

---

†† *Inflated oval, spire short.*

4. **Glandina bullata,** Gould.
Plate 1, figure 5.

Ovate, *ventricose,* finely striate, *very thin, transparent;* spire short, obtuse, suture lightly impressed; whorls 5, slightly convex; body comprising seven-eighths of the shell; aperture two-thirds the total length, lunate, labrum rounded, columella arcuate. Very pale horn color.

Length 37, diam. 20 mill.

Louisiana.

** *Transversely striate, decussated by revolving lines.*

5. **Glandina Vanuxemii,** Lea.
Plate 1, figure 6.

Ovate-fusiform, thin, fragile, translucent; spire moderate, obtuse, apex mamillary, suture crenulated; whorls 7-8, slightly convex; body large, convex, a little inflated; aperture half the length of the shell, nearly three times as long as broad, columella strongly arched. Pale fawn color, sometimes greenish, generally flecked with distant pale spots.

Length 70, diam. 25 mill.

Texas and Mexico.

6. **Glandina decussata**, Deshayes.

Plate 1, figure 7.

Oblong-conic, thin, shining; spire moderate, obtuse, sutures crenulated; whorls 7–8, somewhat convex; body two-thirds the total length, narrowly convex; aperture oblong, one-half the total length, columella curved, covered with a light callus. Light horn color.

Length 50, diam. 18 mill.

Texas, Mexico.

Narrower, more cylindrical than No. 5.

Mexican species.
Not spirally striate.

7. **Glandina turris**, Pfeiffer.

Plate 1, figure 8.

Oblong turrited, thin, diaphanous; spire elevated, obtuse, suture crenulate; whorls 7; body narrow, equalling *three-sevenths of the total length;* aperture semi-oval, columella vertical, abruptly truncate before reaching the base of the aperture. Corneous.

Length 43, diam. 15 mill.

Mazatlan.

8. **Glandina Albersi**, Pfeiffer.

Plate 1, figure 9.

Ovate-oblong, thin, pellucid; spire moderate, conical, apex obtuse, suture scarcely crenulate; whorls $5\frac{1}{2}$–6, slightly convex, the last *scarcely longer than the spire*, attenuated at base; aperture sub-vertical, sinuately semi-oval, labrum slightly arcuate, columella perpendicular. Light corneous.

Length 30, diam. 12 mill.

Mazatlan.

SUCCINIDÆ.

SUCCINEA, Draparnaud.

Shell obliquely ovate, imperforate, very thin, unicolored, corneous, transparent or translucent; spire very small, body whorl large, inflated; aperture large, oval or ovate, peristome simple, acute.

Animal large, with short and thick tentacles; foot broad. Buccal plate with a posterior quadrangular projection for the attachment of its muscles; cutting edge with one or more projections or folds, not striate.

Semi aquatic, inhabiting low, damp ground liable to overflow, on the margins of bodies of water, and frequently adhering to the leaves of aquatic plants. Generally, large numbers of a species are found together—at least of those northern species with the habits of which we are most familiarly acquainted. One species, however, *S. avara*, is very frequently found solitary or in pairs only, and appears to be somewhat different in its habit, as I have frequently taken it in shady places on high ground, far removed from bodies of water.

There appears to be two distinct groups or sub-genera of *Succineæ* inhabiting North America; the first, characterized by well-rounded whorls, is recognized by malacologists as typical; the second, with the shell ovate, and the whorls flattened above and effuse below, has received the name of *Brachyspira*, Pfr.

Examined with a view to geographical distribution, we find the two groups of *Succineæ* to have their representatives in every part of the Union, and yet, so far as we may presume to generalize from the very insufficient data which has been collected and published, the species appear to be much less diffused throughout the country than those of the *Helices*. The following table will give an idea of their territorial range:—

1. Inhabiting the Northern States westward to the Rocky Mountains, extending into British America, southwards to Virginia and Tennessee.
    7, *S. obliqua*, 10, *S. vermeta*, 11, *S. avara*, 21 *S. ovalis*, 32, *S. aurea*.
    a. Confined to New England States.
    1, *S. Tolteniana*.
    b. Confined to northern frontier of the United States and northwards.
    12, *S. Grœnlandica*, 14, *S. Verilli*, 22, *S. Decampii*, 23, *S. Higginsi*.
    c. Ohio.
    24, *S. retusa*.

2. Southern Atlantic States—South Carolina to Florida.
   2, *S. inflata*, 4, *S. campestris*, 6, *S. effusa*, 26, *S. Wilsonii*, 29, *S. luteola*.

3. South-western States—Mississippi, Louisiana and Texas.
   3, *S. unicolor*, 8, *S. Greerii*, 9, *S. Grosvenorii*, 27, *S. Forsheyi*, 28, *S. concordialis*, 31, *S. Salleana*, 33, *S. Haleana*.

4. Pacific States—California, Oregon, British Columbia, Nebraska, etc.
   5, *S. Stretchiana*, 13, *S. Gabbii*, 15, *S. lineata*, 16, *S. Mooresiana*, 17, *S. Oregonensis*, 18, *S. rusticana*, 19, *S. Haydeni*, 20, *S. Sillimani*, 25, *S. Nuttalliana*, 30, *S. Hawkinsii*.

We thus find that our as yet imperfectly-known fauna exhibits thirty-three well characterized species of *Succinea*, and exceeds greatly those inhabiting Europe which, in the last edition of Pfeiffer's Monograph, number thirteen.*

This extraordinary development of species of a single genus of terrestrial Mollusks upon our Continent is the more remarkable from the fact that in other terrestrial genera common to the two Continents, our fauna is by no means so rich as that of Europe. In *Helix*, *Bulimus* and *Pupa*, the species are not nearly so numerous with us.

Several species of *Succinea* inhabit Mexico, Central America and the West India Islands, but their number is not so great as would be expected from the considerable quantity of other types of terrestrial Mollusks in those countries. In fact, the paucity of these forms leads to the conclusion that the maximum development of these groups (*Succinea* and *Brachyspira*) of the genus is within the limits of the United States, and that the principal species are inhabitants of the temperate zone of America and Europe.

* It is remarkable that Pfeiffer and Albers (Helicéen, 1861) have both failed to recognize the true characters of the two groups of *Succinea* inhabiting Europe. Although their definition of *Brachyspira* is sufficiently correct, yet they have so distributed the species as to show that they regarded these divisions as very arbitrary, not as natural ones. Thus, we have almost every species of *Brachyspira*, European and American, placed by these authors in their sub-genus or group *Succinea*, whereas almost every European species is a true *Brachyspira*. We may instance *S. Pfeifferi*, *S. longiscata*, *S. Italica*, etc. *S. Humilis*, *S. Baudoni*, etc., may be classed as typical *Succinea*.

† *Spire short, body whorl very large, inflated, well rounded.*

1. **Succinea Totteniana,** Lea.
Plate 2, figure 1.

Obliquely ovate, thin, transparent, obsoletely striate, shining; spire very short, of scarcely three whorls; body whorl nine-tenths of the total length and inflated oval; aperture oval, obtusely angulate above, three-fourths the total length, peristome well rounded.

Length 16, diam. 9 mill.

New England and Eastern New York.

2. **Succinea inflata,** Lea.
Plate 2, figure 2.

Inflated, oval, thin, spire very short, apex sub-acute, suture impressed; whorls 3, those of the spire not very convex; body short-oval, seven-eighths the total length; aperture widely ovate. Chalky white, spire inclining to brown, aperture within tinged with light yellow.

Length 12, diam. 8 mill.

South Carolina and Georgia.

Is of heavier texture and different color from *S. Totteniana.* Mr. W. G. Binney considers this species doubtfully distinct from *S. Campestris.* I think it is well distinguished.

3. **Succinea unicolor,** Tryon. (Nov. spec.)
Plate 2, figure 3.

Oval, inflated, very thin, translucent, finely striate; spire very short, apex acute, suture moderately impressed; whorls 3, the last very large; aperture short-ovate, outer lip somewhat expanded. Light corneous.

Length 8, diam. 6 mill.

New Orleans, La.

This shell, mentioned by Pfeiffer as a variety, appears to me to be specifically distinct. Specimens exist in the magnificent collection of the Academy of Natural Sciences of Philadelphia, and for others I am indebted to Mr. Thomas Bland, of New York.

4. **Succinea campestris,** Say.
Plate 2, figure 4.

Rounded-ovate, distantly striate, shining; spire short, apex acute; whorls 3, convex, suture impressed; body large, ventricose; aperture oval, not quite two-thirds the total length, columella indented in the middle. Yellowish white or yellowish horn color, the striæ opaque white.

Length 15, diam. 10 mill.

South Carolina to Florida.

Differs from *inflata* in having a somewhat longer, more convex spire, and in the distant, white striæ. In *campestris* the aperture does not occupy so large a portion of the entire width as in *inflata*.

5. **Succinea Stretchiana,** Bland.
Plate 2, figure 5.

Globose-conic, thin, pellucid, shining, striatulate; spire short, obtuse, suture well impressed; whorls 3, convex, the last inflated; aperture roundly oval, columella arcuate, slightly thickened. Greenish horn color.

Length 6·25, diam. 5 mill.

Little Valley, Washoe Co., Nevada, on the eastern slope of the Sierra Nevada, 6500 feet above the sea.

6. **Succinea effusa,** Shuttleworth.
Plate 2, figure 6.

Depressed-oval, very thin, transparent, shining, slightly striated; spire remarkably short, apex acute, body equalling fourteen-fifteenths of the length of the shell; aperture very large, oblique, wide, broadly rounded below, columella scarcely rounded. Greyish horn color.

Length 11, diam. 7 mill.

Florida.

Differs from all the preceding in the minute spire and proportionally very long body, the aperture being four-fifths the total length and two-thirds of the width of the shell.

> ** *Spire moderate, apex acute, body inflated, aperture large, broadly oval, peristome well rounded, the superior part not flattened.*

7. **Succinea obliqua**, Say.
Plate 2, figure 7.

Ovate, very thin and fragile, pellucid, shining, irregularly wrinkled or striate; spire short, minute, suture well impressed; whorls 3, a little oblique, the last very large, expanded, ovate; aperture large, oval, both lips equally rounded, a little angular behind, equalling nearly three-fourths the total length. Yellowish or yellowish green.

Length 20, diam. 13 mill.

British America to Virginia, westward to Arkansas, Iowa.

8. **Succinea Greerii**, Tryon. (Nov. spec.)
Plate 2, figure 8.

Ovate, rather thick, rugose, not transparent; spire short conical, suture moderately impressed; whorls 3, but slightly oblique, well rounded, the last one a little flattened above the periphery; aperture ovate, three-fifths the total length, not acute above, well rounded below; columella twisted, with a slight callous deposit. Light yellowish or greenish horn color.

Length 15, diam. 9 mill.

Vicksburg, Miss.—Col. James Greer.

This species was recently sent to the Academy of Natural Sciences in considerable numbers by Col James Greer. Most of the specimens were smaller than the one figured. It appears to be intermediate between *S. obliqua*, Say, and *S. Grosvenorii*, Lea, differing from the former in color, texture, the contorted and less curved columella, smaller size, and less swollen contour, as well as more regular increase of the whorls; and from the latter by its color, less convexity, more ovate aperture, &c.

9. **Succinea Grosvenorii**, Lea.
Plate 2, figure 9.

Ovate, thin, translucent, shining, distantly striate; spire elevated, apex acute, suture well impressed; whorls 3, a little oblique, the body large but not much inflated; aperture broadly rounded, the columella impressed above, equalling two-thirds the total length. Light lemon color.

Length 12, diam. 8 mill.

Alexandria, La.

Smaller than No. 8, with more elevated spire, more convex whorls, and heavier texture.

10. **Succinea vermeta,** Say.

Plate 2, figure 10.

Ovate-lengthened, very thin, fragile, translucent, wrinkled; spire elevated, acute, suture *profoundly impressed;* volutions 3, *very much rounded,* oblique; aperture ovate, rounded above. Yellowish.

Length 10, diam. 7 mill.

Northern States.

The great convexity of the whorls and very deep suture are the distinguishing characters of this species. A large variety of *S. avara* not possessing these characters has erroneously passed among Conchologists as *S. vermeta.*

11. **Succinea avara,** Say.

Plate 2, figures 11 and 12.

Ovate, thin, fragile, minutely hairy or shining, spire elevated, acute, suture well impressed; aperture broadly oval, *a little more than half the total length.* Yellowish or greyish, *frequently encrusted with dirt.*

Length 6, diam. 3½ mill.

Northern, Middle and Western States to Nebraska.

The smallest of the northern species, readily distinguished from No. 10 by its less scalariform volutions. In Greenland it is replaced by the following species; in the Pacific States by *S. Oregonensis;* in the far South by *S. Haleana*—all species of about equal size. A western variety sometimes attains nearly double the dimensions quoted above.

This species frequently inhabits at a considerable distance from water or low grounds.

Mr. Isaac Lea has described a species from Ohio under the name of *S. Wardiana,* which Mr. Binney believes to be identical with *S. avara,* in which opinion I coincide. As Mr. Lea has recently reasserted the specific weight of his species,[*] I have had a drawing made from the type (fig. 12) of *S. Wardiana,* to afford opportunity of making comparisons, as it had not been previously figured.

[*] Proceedings Acad. Nat. Sciences, 1865.

12. **Succinea Grœnlandica,** Beck.
Plate 2, figure 13.

Elongated, strongly wrinkled; spire elevated, suture well impressed; body very large proportionally, not inflated; aperture oval, three-fifths the total length. Covered with an opaque, somewhat shining, horn-colored epidermis, with narrow white lines caused by its abrasion from the elevated striæ.

Length 8, diam. $5\frac{1}{2}$ mill.

Greenland, and Mingan Island, Gulf of St. Lawrence.

---

13. **Succinea Gabbii,** Tryon. (Nov. sp.)
Plate 2, figure 14.

Elongate ovate, thin, sub-pellucid, coarsely undulately striate; spire long, acute, suture deeply impressed; whorls nearly 4, but slightly oblique, very convex, the last three-fourths of the total length; aperture small, roundly oval, columella well incurved. Light yellowish.

Length 9, diam. 5 mill.

Crooked Creek of Owyhee, 60 miles W. of boundary S. E. Oregon. Crane Lake Valley, N. E. Cal.—W. M. Gabb.

Larger and more convex than *Grœnlandica*, and differently colored. The spire is proportionally longer in this than in the other species of the group.

---

14. **Succinea Verrilli,** Bland.
Plate 2, figure 15.

Ovate conic, thin, sub-pellucid, striate; spire elevated, obtuse, apex globose; suture well impressed; whorls 3, very convex, not very oblique; aperture roundly oval, small, columella arcuate with a slight callus. Orange yellow, apex reddish.

Length 7, diam. 3·5 mill.

Salt Lake, Anticosti Island, Gulf of St. Lawrence.

Smaller than *Grœnlandica*, of different color, and has more distinct and regular incremental striæ.

15. **Succinea lineata,** W. G. Binney.
Plate 2, figure 16.

Oblong ovate, irregularly wrinkled, between which are coarse, remote, revolving lines; spire acute; whorls 3, *very convex;* aperture *one-half* the length of the shell, oval; columella folded.

Length 12, diam. 6 mill.

Nebraska, N. E. California, British America.

Differs from *vermeta* by its more oval form. The aperture is correctly egg-shaped. It is the heaviest of American species. The columella is not always folded, nor the revolving lines apparent.

---

16. **Succinea Mooresiana,** Lea.
Plate 2, figure 17.

Ovate, thin, striate; spire elevated, acute; whorls 3, convex, suture impressed; body large, not inflated; aperture widely ovate, obtusely angled above; columella with a distinct fold. Light yellowish-white.

Length 9, diam. 6 mill.

Platte River.

This may be only a variety of No. 15. The principal difference is, that it is more depressed, the spire not so much exserted and not so convex, and the aperture nearly two-thirds the total length.

---

17. **Succinea Oregonensis,** Lea.
Plate 2, figure 18.

Elongated oval, thin, diaphanous, shining, striate; spire acute, suture well impressed; whorls 3, well rounded; body seven-eighths and aperture two-thirds the total length; aperture ovate, one-third longer than broad, columella arcuate. Color deep orange or golden.

Length 9, diam. 6 mill.

Oregon, California.

Differs from No. 16 principally in its diaphanous texture and dark color.

18. **Succinea rusticana,** Gould.
Plate 2, figure 19.

Elongate ovate, thin, fragile, diaphanous, irregularly striate; spire elevated, acute, suture moderately impressed; whorls 3, not very convex; body long, oval, not inflated; aperture narrowly oval, three-fifths the entire length. Pale greenish or yellowish.

Length 14, diam. 7 mill.

Oregon, California.

Immediately distinguished from the other species of this section, by its narrow, lengthened form. Resembling the following several species, but differing from them all in the regular curve of the outer lip.

*** *Shell ovate, spire rather elevated, apex acute, whorls flattened, body whorl large, but not inflated, aperture angulate above, labrum superiorly more or less flattened.*

(BRACHYSPIRA, Pfeiffer.)

19. **Succinea Haydeni,** W. G. Binney.
Plate 2, figure 20.

Elongate-oval, thin, shining; spire short, acute; whorls 3, convex, the last marked with wrinkles of growth, and irregular, heavy, spiral furrows; aperture oblique, oval, five-sevenths the total length, the lower margin considerably expanded. Amber colored.

Length 21, diam. 9 mill.

Nebraska and northwards.

The peristome is more flexuose than in *S. ovalis*, Gld., and it is more attenuately pointed above; it also differs in having the revolving lines, and is a larger species.

Var. *minor*. Length 15 mill.

---

20. **Succinea Sillimani,** Bland.
Plate 2, figure 21.

Oblong-ovate, thin, coarsely striate, shining; spire short, acute, suture impressed; whorls 3, convex, much flattened superiorly; aperture oblique, elongate oval, angular above, effuse at base, columella slightly arcuate, with a thread-like thickening above. White?

Length 20, diam. 8·5 mill.

>Humboldt Lake, Nevada.

More attenuated than *S. Haydeni*, the last whorl less tumid, and the aperture more narrow.

---

21. **Succinea ovalis,** Gould.
>Plate 2, figure 22.

Ovate-conic, very thin, pellucid, shining, very minutely striate, spire acute, suture slightly impressed; whorls 3, the last compressed and elongate; aperture produced, elongated, broadly rounded below, more than three-fourths the total length. Light yellowish horn color.

Length 12, diam. 6 mill.

New England to Wisconsin and southwards to Maryland and Kentucky.

---

22. **Succinea DeCampii,** Tryon. (Nov. sp.)
>Plate 2, figure 23.

Ovate-conic, rather thick for the genus, translucent, finely striate, surface very much polished; spire short, acute suture moderate; whorls 3, very oblique, narrow, flattened; aperture narrow ovate, columella slightly incurved. Yellowish ash color, spire golden, edge of aperture black.

Length 10, diam. 5 mill.

>Marshall, Michigan. W. H. DeCamp.

Narrower, thicker and more polished than *S. ovalis*. It also differs in color and size. The black-edged peritreme is remarkable, and is present in all the adult specimens I have examined.

---

23. **Succinea Higginsi,** Bland. (Nov. spec.)
>Plate 2, figure 24.

Depressed-oval, thin, pellucid, somewhat shining, pale horn colored; spire short, obtuse, suture deep; whorls 3, convex, the last rather depressed; columella scarcely arched, above conspicuously plicate; aperture angularly oval, frequently armed with a small, oblique, white tooth on the parietal wall; peristome simple, regularly arcuate.

Length 15, diam. 7 mill.

Put-in Bay Island, Lake Erie.

Allied to *S. Salleana*, Pfr., *S. Haydeni*, W. G. Binney, and especially to *S. ovalis*, Gould. Compared with the latter, the last whorl is less convex, the aperture is more angular above, the columella less arcuate and more distinctly plicate. Three specimens had the parietal tooth mentioned in the description. It is the only North American species in which this tooth has been observed.

24. **Succinea retusa,** Lea.

Plate 2, figure 25.

Ovate-oblong, thin and pellucid; spire moderate, acute; aperture two-thirds the total length, elongate ovate, sharply angled above, dilated and retracted below. Light yellowish.

Length 17, diam. 8 mill.

Ohio.

Very close to *ovalis;* rather narrower, and differs in the aperture.

25. **Succinea Nuttalliana,** Lea.

Plate 2, figure 26.

Ovate conic, very thin, pellucid, shining, striate; spire acute, attenuate; whorls revolving very obliquely; aperture two-thirds the total length, ovate, broadly rounded below, angled above; columella without fold. Light horn color or greyish.

Length 15, diam. 8 mill.

Oregon, California.

Aperture slightly narrower posteriorly than *ovalis*. The difference between the two is very slight, but they inhabit different zoological regions. *Nuttalliana* is rather larger than *ovalis*.

26. **Succinea Wilsonii,** Lea.

Plate 2, figure 27.

Elongate-oblique, striate, thin, diaphanous; spire prominent, acute, suture well impressed; whorls 4, rather convex, not very oblique; aperture rather large, ovate, columella slightly incurved and contorted. Orange color.

Length 17, diam. 8 mill.

Darien, Georgia.

---

27. **Succinea Forsheyi,** Lea.
Plate 2, figure 28.

Ovate-conic, striate, thin, diaphanous; spire very short, acute; whorls 3, rapidly increasing, not very oblique; body whorl nearly the entire length of the shell, narrowly oval; mouth oval, a little angled above, columella folded nearly at the superior part of the aperture. Very light lemon color.

Length 11, diam. 6 mill.

Rutersville, Texas.

---

28. **Succinea Concordialis,** Gould.
Plate 2, figure 29.

Ovate-conic, thin, *feebly decussately striate;* spire acute, prominent; whorls rather more than 3, very oblique, rapidly increasing, the upper half of the body whorl flatly compressed; aperture two-thirds the total length, acuminated above, well rounded below; columella with greater curve than the outer lip, slightly angled at its superior termination near the top of the aperture; a thin callus covers the left margin, which is *slightly detached* anteriorly, *forming a rudimentary umbilicus.* Pale honey yellow.

Length 12, diam. 8 mill.

Texas, Mexico.

---

29. **Succinea luteola,** Gould.
Plate 2, figure 30.

Ovate-conic, irregularly wrinkled, *somewhat thickened;* spire moderate, apex acute; whorls 4, those of the spire well rounded; upper half of body obliquely flattened; aperture ovate, over half the total length, columella not folded. Pale yellowish or drab to white, apex and interior deeper yellow.

Length 12, diam. 6 mill.

Florida.

Differs from other shells of same size and proportion in its heavier texture.

30. **Succinea Hawkinsii**, Baird.
Plate 2, figure 31.

Very narrow, sub-cylindrical, thin, rugosely striate; spire very short, apex mamillary; whorls $2\frac{1}{2}$, suture not impressed; body very long and narrow, the sides flattened, sub-parallel; aperture narrow ovate, two-thirds the total length, viewed from the base exhibiting the interior of the whorl to the apex, columella slightly folded above, with a callous deposit. Covered with a rather opaque dark yellow or orange epidermis.

Length 12, diam. 5 mill.

Washington Territory, British Columbia.

No other American species has the peculiar narrow form, fragile substance and opaque epidermis of the above.

31. **Succinea Salleana**, Pfeiffer.
Plate 2, figure 32.

Depressed ovate, somewhat wedge-shaped, very thin, striate, with impressed irregular revolving striae; spire *very* short, not elevated above the general outline of the shell, apex papillary; whorls $2\frac{1}{2}$, very much obliquely flattened above, broadly rounded below; aperture seven-eighths the entire length of the shell, pear-shaped, sharply angled above, columella without fold, not so well rounded as the labrum. Light corneous.

Length 16, diam. 8 mill.

New Orleans.

The narrow wedge-shaped form of this species, together with the spire almost minute and the very long aperture, will amply serve to distinguish it from the others.

32. **Succinea aurea,** Lea.
Plate 2, figure 33.

Very small, elongated oval, very thin, transparent; spire short; whorls 3, a little tabulated posteriorly, suture deeply impressed; aperture narrow-ovate, acutely angled above; columella slightly folded. Amber color.

Length 8, diam. 4 mill.

Ohio, Niagara Falls.

Same size as *S. avara,* Say, but *narrower, more polished and pellucid, and darker color.*

---

33. **Succinea Haleana,** Lea.
Plate 2, figure 34.

Oval, minutely striate; whorls $2\frac{1}{2}$, apex mamillary, suture deeply impressed; body whorl a little flattened around the superior part; aperture widely oval, angled above, columella medially folded, with a slight deposit of callus. Light honey yellow.

Length 5, diam. 3 mill.

Alexandria, La.

Very close to *avara;* the whorls are not so convex, nor the spire so prominent, and the body is proportionally longer. *Avara* does not extend nearly so far southward.

⁂

Mexican Species.

34. **Succinea cingulata,** Forbes.
Plate 2, figure 35.

Oblong-ovate, slightly oblique, striate, shining; spire well developed, suture impressed; whorls 4; aperture large, oval, columella at the base receding to the left. Brownish-yellow, with obsolete spiral white lines.

Length 12, diam. 6 mill.

Mazatlan?

HELICELLIDÆ.

Shell discoidal, orbicular or trochiform, corneous, thin, polished, sometimes transversely striate, translucent or transparent, *lip sharp* (not reflected outwards, nor internally thickened); aperture without proper marginal teeth, but sometimes with internal laminæ not reaching to the edge.

Animal long and narrow. *Buccal plate* thin, crescentic, with an elevation in the middle of the cutting edge, side slightly striate in the centre, or all over.

Lingual dentition.—Uncini long and broad, tridentate, laterals *long, narrow,* curved, bidentate.

Sub-families.

VITRININÆ. *Shell* depressed, very fragile, consisting of about three whorls, the last extremely enlarged; mouth very oblique and large, extending to the centre of the base of the shell.

Animal too large for complete retraction within the shell.

Some of the species of *Vitrina*, as well as *Helix*, have a caudal mucous gland, and would, therefore, in accordance with the views of Gray and others, be placed in another family. We are at present compelled to consider the gland as of *no importance whatever* in classification, or else to construct an exceedingly artificial and unnatural system.

HELICELLINÆ. *Shell* thin, *glabrous*, translucent or transparent, *polished, globosely depressed;* mouth not dentate. Umbilicus generally narrowly perforate. Umbilical region impressed.

Lingual dentition.—As in *Vitrininæ.*

Differs from *Vitrininæ* in the moderate aperture and impressed umbilical region, from *Gastrodontinæ* in being more depressed, and not impressed striate, and from *Patulinæ* in the absence of opaque color, or ribs.

GASTRODONTINÆ.* *Shell* thin, translucent, *striate* or ribbed, generally *depressed conical*, frequently lamellately toothed.

Lingual dentition.—Generally as in the above, sometimes the laterals are *square*, bidentate.

Distinguished from all the others by conical shape, from *Patulinæ*, also, by its narrow umbilicus, and diaphanous texture.

---

* This and the following Sub-family are not proposed with any intention but to facilitate the determination of species. The Sub-family *Vallonina* of Mr. Morse, *in its presnt limits*, we cannot adopt.

PATULINÆ. *Shell* moderately thick, translucent to opaque, epidermis *opaquely* colored, sometimes banded or striped in the large species, *striate* or *ribbed*, discoidal, planospiral, or spire a little globosely elevated. Umbilicus wide. Mouth not toothed, (lamellately toothed in *one* species only.

Lingual dentition.—Uncini bidentate, laterals either long or square, bidentate or tridentate.

VITRININÆ.

Genera.

1. VITRINA, Drap. Shell very thin, polished, transparent, small, consisting of two or three depressed whorls, rapidly increasing. Aperture very large, oblique, lunate. Columellar margin a little inflated. Axis imperforate. Nearly covering the contracted animal.
2. BINNEYA, Cooper. Shell car-shaped, nearly flat, whorls two, last whorl enormously expanded. One-third as long as the animal, which it does not half cover when contracted.

Much more depressed, with a larger proportionate aperture than *Vitrina*.

VITRINA, Drap.

1. **Vitrina limpida,** Gould.

Plate 3, figure 1.

Globosely depressed, whorls $2\frac{1}{2}$, scarcely convex, suture very slightly canaliculate; plane of aperture very oblique. Colorless.

Diam. 6 mill.

Maine to Michigan, and Northward.

2. **Vitrina Angelicæ,** Beck.

Plate 3, figure 2.

Globosely depressed, whorls $3\frac{1}{2}$, suture crenulated, spire small, somewhat prominent; aperture lunately oval, oblique. Color greenish-yellow.

Diam. 6, height, 3·5 mill.

Greenland.

More globose, with one more whorl and more prominent spire than No. 1.

3. Vitrina Pfeifferii, Newcomb.
Plate 3, figure 3.

Globosely depressed, whorls 3; suture very finely margined; aperture large, oblique, roundly ovate; lip thin, columella arched.

Diam. 5, height 2 mill.

Carson Valley, Cal.

V. limpida is smaller, with only 2½ whorls, and thinner. *V. Angelicæ* is more globose, with more prominent spire.

BINNEYA, Cooper.
1. Binneya notabilis, Cooper.
Plate 3, fig. 4.

Depressed, smooth and shining, epidermis extending beyond margin of aperture, translucent when young, but opaquely thickened when old. Nuclear whorl with about thirty delicate transverse ribs. Pale brown.

Diam. 12, height 3 mill.

Santa Barbara I., Cal.

HELICELLINÆ.

Genera.

1. MACROCYCLIS, Beck. Shell moderate, widely umbilicate, planorboid, striate; whorls 4–5, the last wide, descending and *flattened above* at the aperture.

2. HYALINA, Ferussac. Shell moderate or small, globosely depressed, moderately umbilicated or perforated, or umbilicus closed, but impressed; whorls 4–6, vitreous, shining, regularly increasing, not angled at the periphery, nor flatly depressed at the aperture.

MACROCYCLIS, Beck.
1. Macrocyclis Newberryana, W. G. Binney.
Plate 3, figure 5.

Large, whorls 6, first ones flattened, but ultimate one convex, beneath convex; reddish-brown, striate, decussated by fine spiral lines; spire depressed, suture deeply impressed; umbilicus wide and deep, lip margins connected by a callus on the body.

Diam. 37, height 13 mill.

San Diego, Cal.

Differs from all others of the group in its large size, and color. The decussated surface, large umbilicus and rounded lip distinguish it from *M. Vancouverensis*, Lea.

2. **Macrocyclis Vancouverensis,** Lea.
Plate 3, figure 6.

Large, whorls 5, the superior part of the last one flattened upon approaching the aperture, rounded beneath; bright yellowish-green, shining, roughly striate, with very slight revolving lines, suture moderate, umbilicus of moderate width and deep.

Diam. 30 mill.

Oregon and Washington Territory.

3. **Macrocyclis sportella,** Gould.
Plate 3, figure 7.

Medium size, whorls 5, the superior part of the last one flattened upon approaching the aperture, rounded below; very light apple-green, dull, very closely and sharply striate, reticulated by slight, revolving lines; suture moderate, umbilicus moderate and deep.

Diam. 18 mill.

Oregon, California.

With same number of whorls, is much smaller than No. 2, and more sharply striate. Messrs. Binney and Bland consider the two identical, but the differences are permanent in many specimens before me.

4. **Macrocyclis concava,** Say.
Plate 3, figure 8.

Medium, whorls 5, superior part of last one flattened towards the mouth, well rounded beneath; light-horn color or greenish, but almost white; slightly striate, suture well impressed, umbilicus rather wide and deep.

Diam. generally 12 to 15 mill.

Maine to Iowa, southwards to Georgia and Mississippi.

Lighter in color, and much smoother than *sportella*.

5. **Macrocyclis Voyana**, Newcomb.
Plate 3, figure 9.

Small, depressed; whorls 5, convex, the last declining towards the aperture and somewhat flattened or concave above, striate; aperture sinuate above, the lip slightly expanded, its extremities joined by a callus on the body-whorl; below broadly umbilicate. Pale horn-color.

Diam. 12·5 mill.

Canyon Creek, Trinity Co., California.

Smaller, darker colored and of more rugged aspect than the other Californian species; it may also be distinguished by the much greater sinuosity of the upper part of the lip.

6. **Macrocyclis Elliotti**, Redfield.
Plate 3, figure 10.

Whorls 5, depressed, conic or slanting above, suture moderate, striate, polished, well rounded beneath; umbilicus narrow, deep, aperture very oblique, wide; light greenish-yellow.

Diam. 8, height 4 mill.

North Carolina to Georgia.

HYALINA, Ferussac.

1. *Axis deeply indented at base but not perforate.*

1. **Hyalina indentata**, Say.
Plate 3, figure 11.

Whorls 4, flattened, thin, pellucid, polished, corneous, rapidly enlarging; aperture rather large, transverse, the peristome reaching below to the centre of the base of the shell, which is well impresed, but imperforate.

Diameter 5 mill.

From Canada to Florida, and westward to Michigan and Texas.

2. *Perforate or umbilicate.*

* *Very globose.*

2. Hyalina friabilis, Wm. G. Binney.
Plate 3, fig. 12.

Whorls 5, rapidly increasing; shell very globose, thin, polished, faintly striate, suture scarcely impressed; aperture subcircular, sharp lipped, very slightly thickened at base, and a little reflected over the narrow, deep umbilicus; shell very convex below. Color light horn to reddish.

Diameter 20 to 25, alt. 12 to 15 mill.

South Indiana and Illinois, Arkansas, Alabama, Texas.

** *Globosely depressed.*

(Subgenus OMPHALINA, Rafinesque, W. G. Binney.)

† *Closely striate above, smooth beneath.*

$Diam. = 20$ mill.

3. Hyalina lævigata, Pfeiffer.
Plate 3, fig. 13.

Whorls 5, yellowish or fulvous, very closely and regularly striate above, smooth and shining beneath; last whorl expanding towards the aperture, which is rounded lunular; lip simple, slightly reflected around the moderate umbilicus, and much thickened within at base.

Diam. 20 mill.

Ohio, Indiana, and southwards to Florida, Mississippi and Arkansas.

---

†† *Smooth, or coarsely, irregularly striate.*
$Diam. = 25-35$ mill.

4. Hyalina lucubrata, Say.
Plate 3, figure 14.

Depressed, sub-globose; whorls more than 4, much wrinkled, sub-translucent, reddish-brown, polished, beneath paler; umbilicus rather large; aperture nearly orbicular.

Diam. 26, height 12 mill.

Mexico.

Differs in color and in being more depressed and more coarsely striate, from *lævigata*, Pfr.

5. **Hyalina caduca,** Pfeiffer.

Plate 3, figure 15.

Depressed-globose; whorls 5, rapidly increasing, striate, polished, light yellowish with a tinge of green; aperture subrotund, umbilicus moderate.

Diam. 25, height 13 mill.

Texas, Mexico.

Lighter colored than *H. lucubrata,* but doubtfully distinct from it.

---

6. **Hyalina fuliginosa,** Griffith.

Plate 3, figure 16.

Depressed globose; whorls 4½, rapidly increasing, with irregular oblique wrinkles, smooth, shining, suture slightly impressed; aperture transversely subrotund, terminations of margin approaching; umbilicus moderate. Color dark horn or chestnut.

Diam. 25 mill.

Western part of the Atlantic States, Western and Southern States.

Approaches *H. lævigata,* but is smoother, more polished, more depressed, umbilicus larger, aperture more rounded. Differs in color from *H. caduca* and *H. lucubrata.*

---

7. **Hyalina kopnodes,** Wm. G. Binney.

Plate 4, figure 21.

Depressed globose, wrinkled, below smooth, suture moderate; whorls 5, rapidly increasing, sometimes with revolving lines; aperture large, ends of margin approaching; umbilicus small and deep.

Diam. 35, height 13 mill.

Alabama.

Larger than its allies, lighter in color, *more depressed,* and of *heavy texture.*

*** *Shell much depressed.*

††† *Diam.* = 12—16 *mill.*

8. **Hyalina sculptilis,** Bland.
Plate 3, figure 18.

Depressed orbicular, subpellucid, pale horn color above, lighter beneath, regularly striate above and below; whorls 7, planulate, the last rapidly increasing; umbilicus scarcely perforated and almost covered by a reflection of the lip.

Diam. 12½, height 5 mill.

Western North Carolina.

In its depressed form, small size, nearly closed umbilicus and number of whorls, this species is very distinct from any other of the group. The impressed striæ are close and regular.

---

9. **Hyalina cellaria,** Müller.
Plate 3, figure 19.

Much depressed, whorls 5, fragile, polished, very light-greenish above, more thickened and becoming lighter colored below; aperture transverse; umbilicus small and deep.

Diam. 10—12 mill.

Eastern and Middle States near the coast. (Introduced from Europe.)

---

10. **Hyalina inornata,** Say.
Plate 4, figure 22.

Depressed, perforate, smooth, shining, whorls 5, light yellowish horn color, suture moderately impressed; aperture transversely lunar, with a white testaceous internal deposit; lip reaching to the centre beneath; base flattened.

Diam. 16 mill.

Massachusetts to Virginia, and westwards to Iowa, Michigan and Kentucky.

Larger than *H. cellaria*, but with a much smaller umbilicus, being a mere perforation.

11. **Hyalina subplana**, Binney.
Plate 4, figure 23.

Whorls 5½, planulate above and below, brownish, shining, striated near the apex, suture not much impressed; aperture transverse, without calcareous deposit within; umbilical region but slightly impressed, umbilicus very narrow.

Diam. 18 mill.

East Tennessee to Western Pennsylvania.

Differs from *H. cellaria* in having a narrower umbilicus, and from both that and *H. inornata* in the absence of a calcareous deposit on the interior of the aperture at the base; it is darker colored, larger, and has more whorls than either of them, and is more regularly flattened.

----

†††† *Shell small, diameter not exceeding 6 mill., umbilicus generally narrow and deep.*

12. **Hyalina Breweri**, Newcomb.
Plate 4, figure 27.

Discoidal, pale corneous, shining, transparent, suture slightly channeled, broadly umbilicate; whorls 5; aperture lunate, lip thin, simple.

Diam. 5 mill., altitude 2.5 mill.

Lake Taho, California.

Less elevated, more polished, lighter colored, and more openly umbilicate than *H. arborea*.

----

13. **Hyalina nitida**, Müller.
Plate 4, figure 24.

Whorls 4½, depressed, conically sloping above, with well marked suture, convex below; umbilicus moderate, but deep; aperture well rounded. Amber colored.

Diam. 6, altitude 3 mill.

New York to Ohio, and northwards to Great Slave Lake.

More conical and rather larger than *H. arborea* This species was first detected by Dr. Ingalls at Greenwich, New York, who called it *H. hydrophila*. It was subsequently ascertained to be one of the few European species common to both continents. It has been known among American conchologists, until recently, as *H. lucida*, Drap., which is, however, a synonym of *H. nitida*, Müll.

14. **Hyalina arborea,** Say.
Plate 3, figure 17.

Whorls 4½, regularly and moderately increasing, depressed turbinate, thin, amber colored, smooth, shining; lip slightly flexuose; umbilicus moderate and deep.

Diam. 5—6 mill.

Georgia and northwards beyond Canada; westwards to the Rocky Mountains; Los Angelos Co., California.

15. **Hyalina electrina,** Gould.
Plate 4, figure 25.

Whorls 3½, depressed, pale, shining, the last rapidly enlarging towards the mouth; umbilicus very small, deep; lip not flexuous.

Diam. 4 mill.

Maine to Georgia, and westwards to Iowa.

Differs from *H. arborea* in its pale color, more depressed, smaller, fewer whorls, and *their more rapid increase*, and very narrow umbilicus.

16. **Hyalina ottonis,** Pfeiffer.
Plate 4, figure 26.

Orbicularly depressed; whorls 4, very light colored, nearly white, suture narrow, periphery angular; superior surface of whorls obliquely declining, inferior well rounded; umbilicus narrow and deep.

Diam. 5, altitude 2·5 mill.

Florida and West Indies.

Much lighter colored and smaller than *H. arborea*, also angled at the periphery and not so widely umbilicate.

17. **Hyalina vortex,** Pfeiffer.
Plate 4, figure 28.

Very much depressed above, almost planorboid, suture deeply impressed, white, shining; whorls 5, convex beneath; umbilicus narrow and deep.

Diam. 5 mill.

Florida and Cuba.

Has one more whorl, and is more depressed than the other species, also lighter in color.

---

18. **Hyalina capsella,** Gould.
Plate 3, figure 20.

Whorls $6\frac{1}{2}$, planorboid, closely revolving, glistening; amber colored, with distant striæ; aperture narrow, semilunar, extending to the centre of the base, which is minutely perforate.

Diam. 5, altitude 2·5 mill.

East Tennessee.

---

††††† *Shell minute, not exceeding 3 mill.*

19. **Hyalina Binneyana,** Morse.
Plate 4, figure 31.

Whorls 4, spire slightly elevated, pellucid, nearly colorless; aperture well rounded; umbilicus moderate.

Diam. 3, height $1\frac{1}{2}$ mill.

Maine.

Differs from *H. minuscula* in having the spire a little elevated, and from *H. electrina* in being smaller and not increasing the whorls so rapidly. If *Hyalina* is not adopted as a genus, we suggest the name *Helix Morsei* for this species, as *Binneyana* is pre-occupied in that genus.

20. **Hyalina ferrea,** Morse.

Plate 4, figure 32.

Whorls 3, not shining, steel grey; last whorl rapidly enlarging; aperture very large, well rounded; spire slightly elevated, suture distinct, deeply channeled near the apex; umbilicus small and abrupt, exhibiting all the volutions; periostraca minutely marked with fine revolving lines.

Diam. 2·5, height 1·25 mill.

 Maine.

Distinguished from *H. electrina* by having fewer whorls and being smaller in size, also in having a larger umbilicus and revolving striæ.

Mexican Species.

Hyalina bilineata, Pfeiffer.

Plate 4, figure 30.

Whorls 5, shining, light horn color, with a narrow brown band above the periphery and another below the suture; spire slightly elevated, suture not much impressed; umbilicus narrow.

Diam. 15, height 7 mill.

Hyalina zonites, Pfeiffer.

Plate 4, figure 29.

Whorls 6, closely and roughly striate, shining, light horn color, with a narrow, brown band revolving above the periphery, and scarcely concealed above by the spire; spire somewhat elevated, suture distinct; base with the striæ not half so numerous, and much more polished; umbilicus moderate.

Diam. 25, height 12·5 mill.

GASTRODONTINÆ.

Genera.

* *Not dentate.*

1. MESOMPHIX, Rafinesque. Shell moderate in size, umbilicate or perforate; aperture obliquely semilunar.
2. CONULUS, Fitzinger. Shell minutely conical, imperforate or perforate; aperture depressed transverse, its lower margin extending to the basal axis of the shell.

** *Lamellarly dentate.*
† *Outer whorl dentate.*

3. GASTRODONTA, Albers. Shell minute, with one or more laminæ revolving within the base. Surface nearly smooth, polished.

†† *Outer whorl and columella both dentate.*

4. STROBILA, Morse. Shell minute, with laminæ both on the base and the columella. Surface strongly striate.

MESOMPHIX, Raf.

a. *Diam.* = 10—15 mill.

1. Mesomphix intertexta, Binney.

Plate 4, figure 33.

Subglobose, whorls 6—7, closely striated, sometimes with very faint revolving lines, yellowish horn color, slightly angled, and sometimes with a light and beneath it an ill-defined brown band at the periphery. Very convex below. Aperture oblique, narrow lunate, base of shell thickened by a calcareous deposit within; umbilicus small.

Diam. 19 mill.

Western New York to Georgia, and westward to Missouri and Iowa.

Larger than the next, and not quite so conical; the slightly angled periphery and bands of color, when present, offer distinctive characters. The shell is also thicker and not so pellucid, and frequently exhibits spiral lines, not visible on *ligera*.

2. **Mesomphix ligera,** Say.
Plate 4, figure 34.

Subglobose, elevated, obtuse, yellowish horn color, translucent, shining, whorls 6—7, closely striated; aperture obliquely narrowly semilunar, shell thickened within at the base; umbilicus narrow, sometimes closed.

Diam. 12—15 mill.

All the Middle and Western States.

See the distinctive characters under *H. intertexta*.

---

3. **Mesomphix demissa,** Binney.
Plate 4, figure 35.

Depressed convex, whorls 6, shining, yellowish horn-color, thickly, but not coarsely, striate, base rather flattened, umbilicus very small; aperture very oblique, almost transversely compressed-lunate, base of shell thickened within.

Diam. 10—12 mill.

Western Pennsylvania; fossil in Alabama and Texas.

Much more depressed, more solid, and usually smaller than *H. ligera*.

---

b. Diam.=6—7 *mill.*

4. **Mesomphix cerinoidea,** Anthony.
Plate 4, figure 36.

Sub globose, whorls 6, shining, yellowish horn-color, almost smooth, convex below, umbilicus very narrow; aperture semilunar, somewhat oblique, base of shell slightly thickened within.

Diam. 6—7 mill.

North Carolina.

Smoother and more depressed, but otherwise a miniature edition of *M. ligera*.

CONULUS, Fitz.

a. Diam. $= 2\frac{1}{2} - 3$ *mill.*

1. Conulus chersina, Say.
Plate 4, figure 37.

Whorls 5—6, convex, sub-conical, thin, pellucid, smooth, shining, amber-colored, suture well impressed; aperture narrowly transverse, base convex, indented around the closed umbilicus.

Diam. $2\frac{1}{2}$, height 2 mill.

The whole country westward to Rocky Mountains; San Gorgonio Pass, Los Angelos Co., California.

This shell is not the *C. fulva,* of Europe, with which it has been confounded; the differences pointed out by Mr. Morse, in his "Shells of Maine," appear to be constant.

---

2. Conulus Fabricii, Beck.
Plate 4, figure 38.

Whorls 6, convex, sub-conical, apex rather acute, suture profound; whorls striate, narrow, last whorl wider, base convex, impressed at the axis, which is nearly imperforate; aperture transversely lunar; color fulvous, pellucid.

Greenland.

Scarcely distinguished from No. 1 by the sub-perforate umbilicus.

---

3. Conulus Gundlachi, Pfeiffer.
Plate 4, figure 64.

Turbinate, shining, fulvous; whorls 5, convex, the last sub-planulate at base, excavated around the perforation, faintly marked with revolving lines; aperture depressed lunar.

Diam. $2\frac{1}{2}$, height $1\frac{3}{3}$ mill.

Florida, Cuba.

More depressed than No. 2. Differs from No. 1 in the perforate axis.

b. Diam.=1½ mill.

4. Conulus minutissima, Lea.
Plate 4, figure 63.

Globosely turbinate, above obtusely elevated, below convex, fuscous, minutely striate; whorls 4; aperture transversely lunar, umbilicated.

Diam. 1½, height 1 mill.

Maine to Pennsylvania, Ohio.

Distinguished by its small size, striate surface, and well developed umbilical opening.

Mr. Morse has distinguished this species as the genus *Punctum*, (sub-family *Punctinæ*,) from the peculiar conformation of the buccal plate, which is divided into sixteen distinct pieces.

This is the smallest of our species.

GASTRODONTA, Albers.

1. Gastrodonta gularis, Say.
Plate 4, figure 39.

Sub-conical, shining, yellowish horn-color, translucent; whorls 7—8, striate; suture moderately impressed, aperture transverse, rather narrow, margin sharp, extending beneath to the centre of the base, which is barely perforate or closed; within thickened at the base by a callus deposit, and having two parallel lamellar teeth, extending nearly to the basal margin of the lip.

Diam. 7—9, height 5—7 mill.

East Tennessee, Georgia, Alabama.

Sometimes one tooth is wanting.

2. Gastrodonta lasmodon, Phillips.
Plate 4, figure 40.

Depressed orbicular, shining, corneous, translucent; whorls 7, narrow, very slowly increasing, minutely striate, suture moderately marked; aperture nearly circular, laminal teeth upon the internal base of the lip; umbilicus large and deep.

Diam. 6, height 3 mill.

East Tennessee, North Alabama.

Much more depressed than the other species. Is smaller than No. 1, and differs from it in the large umbilicus.

3. Gastrodonta suppressa, Say.
Plate 4, figure 41.

Convexly depressed, thin, spire flattened, shining, pellucid; whorls 6, slowly increasing, minutely striate above, beneath more flattened and smooth, suture moderately marked; aperture transversely semi-circular, callously thickened within at the base, with two parallel lamellæ; umbilicus merely perforated, sometimes covered.

Diam. 6, height 4 mill.

Middle States and Ohio.

Has one whorl less and is more depressed and smaller than No. 1; sometimes there are three instead of two teeth.

4. Gastrodonta interna, Say.
Plate 4, figure 42.

Convexly orbicular, reddish-brown, shining, covered above with close, rounded, very distinct ribs, beneath smooth; whorls 8, narrow, very slowly increasing, suture deeply impressed, periphery slightly angled; aperture transverse, narrow, within thickened, especially at base, with two short lamellæ near the outer portion of the basal margin; margin extending to the axis beneath, which is sometimes narrowly perforate, but frequently closed.

Diam. 6, height 4 mill.

West Pennsylvania to Georgia, and westwards to Missouri.

5. Gastrodonta multidentata, Binney.
Plate 4, figure 43.

Depressed, thin, yellowish horn-color, smooth, shining, pellucid; whorls 6, slowly increasing, suture impressed; aperture transverse, narrow, lip extending to the perforated axis, base convex, thickened within the aperture, through which may be seen two to four rows of 5 or 6 teeth each, radiating from the axis towards the circumference, upon the base of the outer whorl; teeth situated far within, and last row not usually visible from the aperture.

Diam. 3, height 1½ mill.

Green Mountains, Vermont, North-east New York, Maine.

Readily distinguished by the teeth and size of the shell.

STROBILA, Morse.

1. **Strobila labyrinthica,** Say.
Plate 4, figure 44.

Obtuse-conic, brownish; whorls 6, heavily ribbed above, more slightly so beneath, suture well impressed, lip thickened, somewhat reflected, with two revolving laminæ upon the base, not visible from the aperture, but seen through the shell. Upon the body are three revolving laminæ, and on the columella another. Base flattened, umbilicus small, impressed.

Diam. $2\frac{1}{2}$, height $2\frac{1}{2}$ mill.

Maine to Maryland, Mississippi, Texas, Arkansas, Western States.

2. **Strobila Hubbardi,** Brown.
Plate 4, figure 45.

Depressed, thin, striated above, smooth beneath, brownish, thin; whorls 5, with two parietal revolving laminæ, and two more far within on the outer whorl; umbilicus rather wide, lip slightly reflected.

Diam. $2\frac{1}{2}$, height $1\frac{1}{4}$ mill.

Indianola, Texas.

May be distinguished from No. 1 by the teeth being more depressed and having a large umbilicus.

PATULINÆ.
Genera.

1. ANGUISPIRA, Morse. Shell heavy, large, depressed-turbinate, solid, ribbed-striate, banded or striped; umbilicus moderate; aperture not toothed.

2. PATULA, Held. Shell moderate, rather heavy, discoidal, a little convex above, concave below, ribbed-striate, unicolored; umbilicus *very wide* but shallow, exhibiting all the volutions.

3. PLANOGYRA, Morse. Shell minute, perfectly flat above, umbilicus moderate; whorls very convex, the last one crossed by from 20 to 25 sharp raised ribs. Unicolored.

4. HELICODISCUS, Morse. Shell minute, planorboid; whorls equally visible above and below, revolving on the same plane, externally with revolving striæ; aperture lamellarly toothed within the outer lip. Unicolored.

5. PSEUDOHYALINA, Morse. Shell minute, discoidal, slightly convex above, unicolored, closely striate or ribbed; umbilicus large.

Distinguished from *Patula* by the minute size and more moderate umbilicus.

ANGUISPIRA, Morse.

a. Not carinate, sometimes slightly angulate on the periphery.

1. **Anguispira solitaria,** Say.
Plate 4, figure 46.

Globose, thick, coarsely striate; spire turbinately elevated, apex obtuse, suture distinctly impressed; whorls 6, well rounded; body large, well rounded, beneath very convex; aperture sub-circular, the extremities of the lip approaching upon the body; umbilicus large, deep, exhibiting all the volutions. Dark corneous, with (generally two) rufous revolving bands; sometimes nearly white and without bands.

Diam. 25, height 16 mill.

Ohio to Nebraska, and south to Ohio River.

2. **Anguispira Idahoensis,** Newcomb.
Plate 4, figure 54.

Turbinately conic, ashy horn-color; apex obtuse; whorls 5, very convex, the first nearly smooth, the others strongly transversely ribbed, ribs on last whorl numbering 20—26; aperture circular, very oblique; deeply and moderately umbilicate.

Diam. 13, altitude 11 mill.

Idaho Territory.

Distinguished from all the other species by its more conical form, and sharp, distinct, distant ribs.

3. **Anguispira Cooperii,** W. G. Binney.
Plate 4, figure 52.

Globosely elevated, solid, obliquely roughly striated, intersected by delicate, spiral lines; spire elevated, obtuse, suture deeply impressed; whorls 5, convex; body very convex, deflected at the aperture; aperture very oblique, circular, extrem ities of margin nearly joining, connected by a heavy callus; umbilicus moderate, deep. White variously marked with a single narrow, brown band, or two bands, or broader longitudinal and spiral patches.

Diam. 15—25, height 9—12 mill.

Nebraska, Washington Territory.

Smaller than No. 1, with rougher striæ, and *revolving lines;* the umbilicus is also proportionally smaller. In some specimens the spire is more flattened.

---

4. **Anguispira alternata,** Say.

Plate 4, figure 47.

Convex, more or less elevated, obliquely closely ridge-striated; spire slightly or considerably elevated, suture well marked; whorls 6, moderately increasing, not very convex; body moderate, very convex beneath, often slightly angled at the periphery; aperture ⅜ths circular, oblique; umbilicus large and deep, exhibiting the volutions. Light corneous, variegated by oblique irregular brown stripes or spots above and below.

Diam. 20-25, height 8-10 mill.

Whole country eastward of Rocky Mountains.

Varies considerably in ornamentation, convexity of the upper surface, and prominence of the rib-like striæ.

Var. Fergusonii, Bland. Smooth, never carinated.

Diam. 15, height 6 mill.

New York, New Jersey.

Var. alba, Tryon. Perfectly colorless.

Maine, (Morse.) Michigan, (Currier.)

---

b. Carinate.

5. **Anguispira strigosa,** Gould.

Plate 4, fig. 49.

Depressed orbicular, thick, striate; spire not much elevated, flattened, suture impressed; whorls 5; body moderately large, angulate at periphery, strongly deflected at aperture; aperture obliquely circular, lip very nearly continuous; umbilicus wide and deep. Ash-grey to brown, with generally a faint medial band, and numerous bands beneath it.

Diam. 20-25, height 10-12 mill.

Washington Territory, Oregon, Nebraska, N. Mexico.

6. **Anguispira Cumberlandiana,** Lea.
Plate 4, figure 48.

Lenticular, acutely carinated, thin, coarsely ribbed, striate; spire convex, much depressed, suture not prominent; whorls 5, slowly increasing, margined by a carina; aperture somewhat rhomboidal; umbilicus broad and deep. Pale yellowish or ash-color, with irregular transverse brown blotches.

Diam. 13–18, height 5–6 mill.

E. Tennessee.

Very close in color and striation to *alternata*, but differs entirely in its very depressed, lenticular form and very acute carina.

PATULA, Held.

1. **Patula perspectiva,** Say.
Plate 4, figure 50.

Nearly discoidal, slightly convex above, and concave below, strongly striate; whorls 6, suture deeply impressed; aperture small, ⅞ths rounded, generally in adult shells with a very slight tubular thickening (scarcely a *tooth*) within the base; umbilicus very wide, cup-shaped, shallow, exhibiting all the volutions. Corneous, reddish-brown.

Diam. 9, height 3 mill.

W. New York to N. Georgia, westward to Arkansas and Michigan.

The *tooth* described by Binney appears to be an imperfectly developed *fulcrum*.

---

2. **Patula striatella,** Anth.
Plate 4, figure 51.

Depressed convex, nearly discoidal; whorls less than 4, with delicate oblique striæ; suture distinct; aperture rounded, transverse; umbilicus very large, shallow. Light horn-color.

Diam. 5, height 2½ mill.

Maine to Great Slave Lake, B. A., southwards through W. New York and Pennsylvania to Ohio River, and westward to Kansas; District of Columbia.

Much smaller, with fewer whorls, and more elevated than No. 1. In the New England States it entirely replaces *S. perspectiva*.

3. **Patula Durantii,** Newcomb.
Plate 4, figure 53.

Depressed, discoidal, opaque, very minutely striated; *spire not at all elevated*, perfectly plane above; whorls 4, the last shelving; suture linear; aperture rounded, lunate, ends of lip margin approaching; broadly and perspectively umbilicated. Pale corneous.

Diam. 5, height $1\frac{3}{4}$ mill.

Santa Barbara Island, Cal.

---

4. **Patula Whitneyi,** Newcomb.

Nearly flat above, *smooth*, suture well impressed; whorls 4; aperture lunate; with a perspective umbilicus. Smoky horn-color.

Diam. 5, height $2\frac{1}{2}$ mill.

Sierra Nevada, Cal.

---

5. **Patula Cronkheitei,** Newcomb.

Somewhat depressed, a little convex above, *ribbed-striate;* whorls 4, suture wide and deep, almost channeled; aperture rounded; umbilicus large, somewhat perspective. Yellowish horn-color.

Diam. 5, height $3\frac{3}{4}$ mill.

Klamath Valley, Oregon.

More elevated and more strongly striate and sutured than *striatella*, Anth.

PLANOGYRA, Morse.

1. **Planogyra asteriscus,** Morse.
Plate 4, figure 55.

Elevated, planorboid; whorls 4, very convex; suture deep; surface with 25-30 very oblique, thin, raised ribs, between which it is finely striate; umbilicus moderately large, showing all the volutions. Light brown.

Diam $1\frac{1}{2}$, height $\frac{3}{4}$ mill.

Maine, Massachusetts.

Differs from *exigua*, Stimp., by being smaller, the spire not elevated, and ribs not so numerous.

HELICODISCUS, Morse.
1. Helicodiscus lineata, Say.
Plate 4, figure 60.

Discoidal, greenish-yellow; whorls 4, visible below as well as above, with numerous parallel revolving lines, suture well impressed; aperture narrow-lunate; base shallow-concave; a pair of teeth within the outer lip, remote from the margin, and another pair further within and visible through the translucent periphery, in each pair one being placed above, the other below it.

Diam. 3 mill.

Maine to Virginia, westwards to Ohio, Texas.

PSEUDOHYALINA, Morse.
* *Diam.* = 5 *mill.*
1. Pseudohyalina limatula, Ward.
Plate 4, figure 65.

Almost planorboid; whorls 4½, increasing regularly, well rounded; suture very distinctly impressed; aperture small, almost round; umbilicus rather large, as wide as the last whorl, well defined and deep. Color very light, nearly white.

Diam. 5 mill.

New York, Ohio, Indiana, Michigan.

** *Diam.* = 2½ *mill.*
2. Pseudohyalina minuscula, Binney.
Plate 4, figure 62.

Whorls 4, depressed, whitish, slowly increasing in diameter; suture deep; aperture sub-rotund; umbilicus large and deep.

Diam. 2,—2½ mill.

United States east of Rocky Mountains, from Maine to Florida, and West Indies.

3. **Pseudohyalina incrustata,** Pocy.
Plate 4, figure 61.

Depressed, spire slightly elevated, suture deep; whorls 4-5, well rounded, slowly increasing; mouth expanding, nearly circular, the ends of the lip-margin closely approaching and united by a callus; umbilicus one-third the entire diameter, showing all the whorls. Brown, with a ferruginous deposit.

Diam. 3, height 1½ mill.

Texas (from Cuba.)

---

4. **Pseudohyalina conspecta,** Bland.
Plate 4, figure 58.

Umbilicate, sub-depressed, thin, with oblique, rather distant, rib-like striæ; dark horn color; spire convex, apex obtuse, smooth, suture deep; whorls 4, convex, gradually increasing, slightly descending towards the mouth; aperture oblique, lunate-rounded, margins approaching.

Diam. 2, altitude 1 mill.

San Francisco, California.

The spire is more raised, and the ribs more numerous and not so prominent as in *H. asteriscus*. It is distinguished from *H. Mazatlanica*, by its more distinct ribs and smaller umbilicus.

*** $Diam. = 1$-$1\frac{1}{2}$ mill.

5. **Pseudohyalina exigua,** Stimpson.
Plate 4, figure 57.

Discoidal, a little convex above, suture moderate; whorls 3½, spirally striate, with oblique transverse ribs; aperture rounded; umbilicus wide, shallow, exhibiting the volutions.

Diam. 1¾ mill.

Canada, Massachusetts, Minnesota, around Lake Superior.

---

6. **Pseudohyalina millium,** Morse.
Plate 4, figure 56.

Depressed, convex above, transparent, shining, distinctly and regularly striate above, with microscopic revolving lines, more apparent beneath; whorls convex, rapidly enlarging; suture very deeply impressed; umbilicus quite large and deep, exhibiting all the volutions. White, with a greenish tinge.

Diam. 1½, height ½ mill.

Maine.

***

Mexican Species.

Pseudohyalina Mazatlanica, Pfeiffer.

Plate 4, figure 59.

Depressed, umbilicate, costato-striate, corneous; whorls 4, somewhat convex, the spire slightly raised; last whorl narrow, scarcely descending at the aperture; umbilicus equalling $\frac{1}{3}$ of the diameter; aperture oblique, lunately rounded, its margins approaching.

Diam. 2·3, altitude 1 mill.

Mazatlan, Mexico.

Family HELICIDÆ.

Shell depressed, or globosely elevated, strong, striate, with the epidermis colored, frequently banded, opaque, lip either margined within, or expanded, or appressed and reflected. Aperture sometimes toothed.

Animal snail-like, not so narrowly lengthened generally as in *Helicellidæ*. *Buccal plate* arcuate, thick, with transverse rounded ribs.

Lingual dentition.—Uncini and laterals the same in form, the former 1–2 dentate or notched irregularly, the latter 1 dentate.

Subfamilies.

HYGROMIINÆ.—*Shell* not toothed, lip not reflected, sometimes expanded, more or less thickened within the margin.

MESODONTINÆ.—*Shell* frequently toothed, lip broadly reflected and appressed.

HYGROMIINÆ.

Genera.

* *Umbilicate.*

1. HYGROMIA, Risso. Globosely depressed, not angulated, generally hirsute; whorls 5–7, convex; aperture rounded or widely lunate; lip acute, slightly expanded and thickened within. Corneous, generally unicolored. Size small.

    Inhabits east of Rocky Mountains.

2. AGLAJA, Albers. Depressed-conoidal, sometimes obscurely angulate; lip thickened within, encroaching a little on the umbilicus. Yellowish-brown, almost always banded. Size large; surface malleate.

    Inhabits California and Oregon.

3. ARIONTA, Leach. Globosely turbinate; lip thickened within, expanded, dilated at the base so as nearly to cover the umbilicus. Color yellowish-brown, banded. Size large.

    Inhabits California and Oregon.

4. POLYMITA, Beck. Globosely turbinate; lip much thickened within; columella *diagonal, much thickened, and frequently bearing a lamelliform or rounded tubercle;* umbilicus almost entirely covered. Shell large, thick, flesh-color, with generally several revolving bands of darker colors.

Inhabits West Indies, Mexico, Southern California.

   ** *Shell imperforate.*

5. TACHEA, Leach. Shell imperforate, turbinate or depressed, upper whorls flattened, last one convex, descending obliquely to the mouth, which is obliquely semicircular; peristome expanded, within labiate, expanded and appressed into and completely covering the umbilicus. Size moderate. Yellowish, more or less numerously banded.

European, introduced into the seaports and islands of the Eastern States.

6. POMATIA, Beck. Globular, large, last whorl very large, ventricose, deflexed at the aperture, which is orbicularly lunate; peristome slightly thickened within, reflexed and appressed over the umbilicus. Light horn-color, banded.

European, introduced into the seaports of the Southern States and West Indies.

HYGROMIA, Risso.

The five species here united, evidently constitute two distinct groups; the two first species being of European origin, while the other three belong to the Territories bordering on the Gulf of Mexico: yet we can find no characters of sufficient importance to justify their division into two generic groups.

HYGROMIA, Risso.

1. **Hygromia rufescens**, Pennant.
Plate 5, figure 1.

Depressed, subglobose, subangulate; spire depressed conical; whorls 6, somewhat convex, brownish, the last with a white band on the angulate periphery; not descending at the aperture, which is ovately lunar, slightly reflected over the rather large umbilicus.

Diam. 11, altitude 6 mill.

Montreal, Canada East, (J. F. Whiteaves.)

A common European species, introduced as above.

---

2. **Hygromia hispida**, Linnæus.
Plate 5, figure 2.

Rather depressed, moderately umbilicate, corneous, covered with short, hispid hairs; whorls 5-6, somewhat convex, narrow, slowly increasing; aperture semilunar, labiate within.

Diam. 10, altitude 5½ mill.

Canada, Nova Scotia, Massachusetts. (Introduced from Europe.)

H. porcina, Say, a species described, evidently, from immature specimens, is now referred to this species.

---

3. **Hygromia jejuna**, Say.
Plate 5, figure 3.

Subglobose, spire prominent, suture impressed; whorls 5, the last ample, striæ scarcely visible; mouth moderate, semicircular; lip expanded, white, (the whorl grooved behind it,) internally ribbed or margined; umbilicus small, base of shell convex. Light yellow, sparingly hirsute.

Diam. 8, height 6 mill.

Georgia, Florida, Alabama.

I agree with Messrs. Binney and Bland in regarding this species, described from immature specimens, by Say, as identical with *H. Mobiliana*, Lea. Mr. Lea states that *H. Mobiliana* has a *reflected* lip, which at first sight, mature specimens *do* appear to have, caused by the external constriction, and the great thickening within.

4. **Hygromia Berlandieriana**, Moricand.

Plate 5, figure 4.

Globose, spire elevated, prominent, suture deeply impressed; whorls 5, well rounded, thin, translucent, scarcely striate, broadly rounded at periphery, contracted around the aperture; lip much expanded, white, with a much thickened internal margin; parietal wall sometimes with a deposit of callus; base rounded, umbilicus minute. White to yellowish-green, with sometimes a faint, narrow brown band above the periphery.

Diam. 12, height 9 mill.

Arkansas to Texas, Mexico.

Judging from the figure of *H. virginalis*, Jan., published in Chemnitz "Conchylien Cabinet." I do not agree with Binney in considering that species a synonym of *Berlandieriana*, but believe it to = *griseola*.

5. **Hygromia griseola**, Pfeiffer.

Plate 5, figure 5.

Globosely depressed, spire convexly elevated; whorls 4–4½, well rounded, slightly striate; aperture lunar, lip white, a little expanded; umbilicus very narrow. Light brown, with a darker band, bordered with white, above the periphery.

Diam. 10, altitude 6 mill.

Texas, Mexico, Guatemala.

Smaller, more depressed, and differently colored from *H. Berlandieriana*.

AGLAJA, Albers.

This group includes most of the large, brilliantly colored Californian species, and is remarkably restricted, none of its members inhabiting east of the Rocky Mountains. Albers places most of those known to him in the genus *Arionta*, leaving only one—*A. fidelis*, in *Aglaja*.. An examination of the West Coast species of *Arionta*, in Albers, shows that he has confounded, in that genus, two distinct groups, of which, that containing the more globose species, with nearly covered umbilicus, really pertains to it; while those that are turbinately depressed, belong to *Aglaja*. Very many of these shells have never been figured, and very little is known regarding them. I have supplied figures from type specimens, wherever possible, and hope at a future time to be able to complete their illustration.*

* *Hirsute, subangulate at the periphery.*
† *Nearly black, not banded.*

1. **Aglaja infumata,** Gould.
Plate 5, figure 6.

Large, solid, depressed-trochiform, angulate at periphery, suture not much impressed; whorls 6½, not very convex, closely irregularly rugose-striate, granulate and hirsute; aperture rhomboidal, lip very slightly thickened within, scarcely expanded; base convex, umbilicus narrow. Brown, nearly black; aperture shining, chocolate within.

Diam. 37, height 16 mill.

From Humboldt Bay to San Pablo Bay, Cal.

†† *Light brown, with a brown band, bordered with white on each side.*

2. **Aglaja Hillebrandi,** Newcomb.
Plate 5, figure 7.

Depressed-trochiform; spire not much elevated, apex obtuse, suture moderate; whorls 6, very slightly convex, rather flattened, a little descending at the aperture, finely striate, hirsute; periphery angulated; aperture widely lunate, lip expanded; reflected below, thickened within, umbilicus moderate. Yellowish horn-color, the periphery with a red band, bordered with white on either side.

Diam. 22, height 9 mill.

Tuolumne Co., Mariposa, California.

* For particular information regarding the geographical distribution of the Californian Helices, see a paper by Dr. Wesley Newcomb, in American Journal of Conchology, I, p. 342, Oct., 1865.

** *Not hirsute, globosely turbinate.*
† *Nearly black, with sometimes a red band.*

3. **Aglaja fidelis,** Gray.
Plate 5, figure 8.

Subconical, moderately elevated; spire depressed-trochiform; whorls 7, rounded; suture distinct; surface thick, rugosely striate, with slight impressed revolving striæ; aperture obliquely semilunar, lip a little reflected below, scarcely thickened within, partially covering the umbilicus. Light brown to black, with generally a narrow red band, chocolate within, lip pink.

Diam. 37, height 20 mill.

Oregon.

†† *Reddish-ashen, not banded.*

4. **Aglaja anachoreta,** W. G. Binney.
Plate 5, figure 9.

Orbicularly convex; spire elevated, conic, suture impressed; whorls 6, granulated; aperture transversely rounded, lip thickened, slightly expanded, the extremities approaching, partly covering the umbilicus. Reddish-ashen, lip tinged with violet.

Diam. 26, height 14 mill.

California.

Is this a variety of the following species? It is very like it in form, but has no band.

††† *Yellowish-brown, with a narrow dark band.*

5. **Aglaja arrosa,** Gould.
Plate 5, figure 10.

Globosly conic, spire elevated, suture not much impressed; whorls 7, somewhat convex, declining a little at the aperture, rugosely striate, malleate, with indistinct revolving lines; aperture widely semilunar; lip narrowly expanded, a little reflected below; umbilicus partly covered. Brown, with a dark brown, nearly black band above the periphery, visible on the spire; light chocolate within.

Diam. 35, height 20 mill.

San Pablo Bay to Bay of Monterey, California.

6. Aglaja exarata, Pfeiffer.
Plate 5, figure 11.

Depressed-conical; spire short, conical, apex acute; whorls 7, rugose, malleated, slightly convex, the last wide, slightly descending at the aperture; aperture oblique, widely lunar, lip a little thickened, white, a little reflected below; umbilicus moderate; brownish, with a chestnut band.

Diam. 30, height 16 mill.

California.

Dr. Newcomb mentions a variety of a creamy-white color, without bands.

---

7. Aglaja Ayresiana, Newcomb.

Rounded-trochiform; whorls 7, slowly increasing, convex, the first ones superiorly with many rib-like striæ, and numerous spiral lines, deeply impressed; inferiorly pale, and with, minute decussating striations; suture well marked; aperture roundly ovate; lip a little expanded; umbilicus partly closed. Yellowish-white, with a broad black band.

Diam. 22, height 15 mill.

Nootka Sound, Northern Oregon.

---

8. Aglaja Nickliniana, Lea.
Plate 5, figure 12.

Conic-globose, rather thin; spire elevated; whorls 6, moderately convex. the last ventricose, finely granulated, polished; aperture rounded, forming two-thirds of a circle, lip a little expanded above, more so below; base depressed at centre, the umbilicus small and party covered by the lip. Light yellowish-brown, with a brown band.

Diam. 21, height 18 mill.

California.

Dr. Newcomb believes *H. redemita* to be a variety of this species.

9. **Aglaja Carpenterii,** Newcomb.

Roundly conical; apex obtuse; whorls 5½, convex, strongly striated, with numerous fine spiral lines; suture well impressed; aperture circular, margins approaching; lip a little expanded; umbilicus open. Brownish, with an obscure dark band.

Diam. 23, height 16½ mill.

Tulare Valley, California.

Distinguished by its rounded aperture.

*** *Not hirsute, malleated, globosely depressed, not turbinate above.*

a. *With a brown band.*

10. **Aglaja tudicolata,** Binney.
Plate 5, figure 13.

Convexly orbicular; spire depressed-conic; whorls 5, a little convex; body large, expanding somewhat towards the aperture, obliquely wrinkled, malleated; aperture transverse, rather circular; lip a little expanded, reflected nearly, sometimes entirely, over the small umbilicus; base convex. Light yellowish-green or brown, with a broad dark band above the periphery, margined with white. Band visible on the spire.

Diam. 31, height 22 mill.

California.

---

11. **Aglaja Bridgesii,** Newcomb.

Depressed-globose; spire conical, suture well impressed, whorls 6, convex, plicately striate and minutely granulate; aperture round-lunate; lip expanded, reflected below, umbilicate. Translucent, grayish horn-color, with a narrow brown band.

Diam. 27, height 19 mill.

San Pablo, California.

Not solid, larger than *ramentosa*, Gould, spire more elevated, darker in color, and less granulated.

12. Aglaja mormonum, Pfeiffer.
Plate 5, figure 14.

Globosely depressed, thin, arcuately striate; spire slightly elevated; whorls 6, slightly convex, the last descending a little at the aperture; aperture obliquely lunar; lip white margined, the extremities converging, expanded, reflected towards the base; base convex, umbilicus moderate. Light reddish-brown, with a darker band above the periphery, margined with white.

Diam. 29, height $12\frac{1}{2}$ mill.

California.

---

13. Aglaja ramentosa, Gould.
Plate 5, figure 15.

Depressed-orbicular, thin, granulated; whorls $5\frac{1}{2}$, the last a little obtusely angulated; aperture obliquely ovate-oblong; lip white, slightly expanded above, reflected below; perforate. Brownish, with a peripherical band of dark brown, margined with white.

Diam. 20, height $12\frac{1}{2}$ mill.

California.

According to Dr. Newcomb, (Am. Journ. Conch., i., p. 344,) *H. reticulata*, Pfeiffer, is a synonym of this species. I have not seen an authentic specimen of *H. reticulata*, but give a figure copied from a wood-cut loaned to me by Thomas Bland, (*Plate* 6, *fig.* 18.)

---

14. Aglaja Traskii, Newcomb.
Plate 5, figure 16.

Depressed-globose, thin; spire subplanulate; whorls 6, the last *not descending*, with numerous microscopical interwoven striæ; lip but little thickened. Pale corneous, somewhat pellucid, brown banded, within tinged with purple.

Diam. 26, height 16 mill.

Los Angelos, Cal., Santa Barbara.

Differs from the following in its lighter substance and color, the lip not so much thickened, and the body-whorl not descending at the aperture.

15. **Aglaja Dupetithouarsii,** Deshayes.
Plate 5, figure 17.

Orbicularly convex, moderately thick, smooth or substriate; spire obtusely conoidal; whorls 7-8, narrow, the last inflated; aperture ovately semilunar, lip a little expanded; umbilicus moderate. Dark chocolate or light greenish when perfectly fresh, with a dark narrow band above the periphery, margined with white, band visible on the spire.

Diam. 29, height 17 mill.

Near Monterey, California.

Closely allied to *H. mormonum,* but more elevated.

---

16. **Aglaja rufocincta,** Newcomb.
Plate 6, figure 20.

Depressed-globose, with impressed suture; whorls 6, the last *not descending* at the aperture, minutely decussately striate; aperture subrotund, lip expanded, columella not callous; umbilicate. Horny, red banded, lip white.

Diam. 17, height 10 mill.

San Diego and I. Santa Catalina, Cal.

---

17. **Aglaja Gabbii,** Newcomb.
Plate 6, figure 19.

Depressed-globose; spire convex, suture well impressed; whorls 5, convex, the last descending; aperture suborbicular, lip white, expanded, umbilicus very small, partly covered. Pale corneous, with an indistinct brown band.

Diam. 10, height 5 mill.

San Clemente I., Cal.

†† *Not landed.*

18. **Aglaja Rowellii**, Newcomb.

Depressed-globose; spire but little elevated, projecting at apex like a nipple, suture moderate; whorls 4½, polished, very finely obliquely convex, the last large, descending at the aperture; aperture circular, lip thin, a little expanded, margins continued, adhering to the last whorl; umbilicus open. Opaque-white (bleached?)

Diam. 15, height 10 mill.

Arizona.

I have not seen this species, but doubt (from the description) whether its affinities are with this group.

ARIONTA, Leach.

The American species are peculiar to Southern California and Northern Mexico. The type of this genus, *A. arbustorum*, (as well as several other species,) is European, but the following are scarcely distinguishable from it, although so widely asunder in distribution. As already stated, many of the species placed in *Arionta* by Albers, really belong to *Aglaja*; *H. bicincta*, Pfeiffer, and *H. Townsendiana*, are also erroneously classed here by Albers.

1. **Arionta Veitchii**, Newcomb.
Plate 5, figure 19.

Subglobose; spire turbinate, elevated, suture not very distinct; whorls 6, a little convex, the last very large, declining towards the aperture; rather thin, obliquely striate, and sometimes a little spirally corrugated; aperture subcircular, lip a little expanded, and very slightly thickened, partly reflected over the narrow umbilicus; base of shell very convex. Yellowish-white, with numerous irregular, interrupted, revolving brown bands.

Diam. 23, height 19 mill.

Cerros Isle, Lower Cal.

2. **Arionta Californiensis,** Lea.
Plate 5. figure 20.

Subglobular, thin, transparent, slightly granulated and striate, shining; spire elevated; whorls 5, convex, the last very broad; aperture rather small, subcircular, lip slightly everted, thickened within, at the base more reflected, nearly covering the small umbilicus. Pale yellowish horn-color, minutely flecked with pale spots, with a narrow brown, pale margined band above the periphery.

Diam. 18, height 15 mill.

California.

---

3. **Arionta Kelletti,** Forbes.
Plate 6, figure 1.

Depressed-globose, thin, rugose-granulated; spire subturbinate, suture moderate; whorls 6, a little convex, the last large and well rounded at base; aperture wide-lunate, lip expanded, partly covering the narrow umbilicus. Reddish-brown, with a darker band on the spire and a lighter one on the periphery of the last whorl.

Diam. 22, height 19 mill.

San Diego, Cal.

---

4. **Arionta crebristriata,** Newcomb.
Plate 6, figure 2.

Moderately thick, depressed-globose; spire turbinate, suture well impressed; whorls 5, a little convex, the last descending towards the aperture, with dense, strong, transverse and minute, longitudinal striæ; aperture rounded, lip either thin, acute, or thickened within, its extremities approaching, sometimes connected by a callus; umbilicus partly covered by the lip. Dark horn-color, obsoletely banded, livid within the aperture.

Diam. 23, height 14–20 mill.

San Clemente I., Cal.

Variable in elevation, and in the thickening of the lip. Differs from *Kelletti*, Forbes, in sculpture.

**
Mexican Species.

5. **Arionta Rémondi,** Tryon.
Plate 5, fig. 18.

Turbinately globose, very thin; whorls 4, scarcely striate, (punctate when viewed with a lens,) slightly convex, the last large, rounded; base convex; umbilicus narrow, with an angled margin; aperture obliquely semilunar, lip expanded. Light corneous, with a narrow brown band on the periphery, and above the suture on the spire.

Diam. 17, height 12 mill.

Cinaloa, near Mazatlan.

---

6. **Arionta Humboldtiana,** Valenciennes.
Plate 6, figure 17.

Ventricose, roughly irregularly striate and wrinkled, malleated; spire small, acuminate; whorls 4, rapidly enlarging, the last very large; aperture oblique, large, lip expanded, its extremities connected by a thin testaceous deposit; umbilicus partly covered. Grayish-white, with three rufous bands on and above the periphery.

Diam. 37, height 28 mill.

Mexico.

I include this species because it was figured by Dr. Binney in his Terrestrial Mollusks, by error, as *Pomatia aspersa.* It does not even belong to the same genus, although placed there by Albers.

POLYMITA, Beck.

This group includes, according to Albers, a large collection of West Indian species, to which we now add several Californian forms.

1. **Polymita Tryonii**, Newcomb.
Plate 6, figure 3.

Solid, depressly globose; spire subturbinate, obtuse, suture well impressed; whorls 6, convex, with numerous minute revolving lines; aperture rounded, lip scarcely expanded, thickened within; columella diagonal, with one or two obsolete tooth-like prominences; umbilicus narrow, covered. Ashy sky-blue above, mottled by streaks of brown, yellowish-white below, an indistinct brown band on the periphery.

Diam. 25, height 19 mill.

Santa Barbara and San Nicholas Isles, Cal.

Var. Superior whorls with white transverse undulating lines.

---

2. **Polymita intercisa**, Wm. G. Binney.
Plate 6, figure 4.

Solid, globose-depressed; spire conic; whorls 5, slightly rounded, closely deeply *striate, crossed by deep revolving lines;* aperture obliquely lunar, lip heavy, thickened, dirty white, the extremities connected by a heavy ash-colored callus; umbilicus covered by the lip, which internally at the base is furnished with a tooth-like process or elevation. Grayish-yellow, with an obscure brown band.

Diam. 22, height 15 mill.

Oregon.

---

3. **Polymita areolata**, Sowerby.
Plate 6, figure 5.

Globose-conic, thin, striate; spire depressed-conoidal; whorls 5, a little convex, the last slightly descending towards the aperture, large, inflated; aperture rounded lunate, lip thickened within, columellar margin sometimes somewhat dentate, nearly covering the narrow umbilicus; base convex. Cream-color, ornamented with revolving series of interrupted brown lines, light brownish or reddish within.

Diam. 26, height 18 mill.

Oregon, California.

4. **Polymita redemita,** Wm. G. Binney.
Plate 6, figure 7.

Globose-conic, rather thin, wrinkled, closely and minutely granulate; spire elevated, obtuse, suture impressed; whorls 6, convex, last quite large and rounded, depressed towards the aperture; aperture rather large, very oblique, lip reddish ash-color, thickened, ends approaching, entirely covering the umbilicus. Brown, banded with chestnut above the middle.

Diam. 21, height 12 mill.

California.

Dr. Newcomb considers this a variety of *Nickliniana*, but it appears to me that it is distinguished by its closed umbilicus, as well as by texture and color.

---

5. **Polymita pandoræ,** Forbes.
Plate 6, figure 8.

Depressed-globose, thin, wrinkled, minutely striate; whorls 5, the last descending towards the aperture; aperture rounded, the lip thickened internally, expanded, dilated and reflected, covering the umbilicus. Brown or violet above, whitish beneath, the periphery encircled by a brown band, brown within, with a white band.

Diam. 17, height 14 mill.

Santa Barbara, and Southern California.

This species is smaller, but very closely allied to *P. Tryonii*.

---

6. **Polymita levis,** Pfeiffer.
Plate 5, figure 21.
Plate 6, figure 6.

Globose, thin, obliquely striate, obsoletely granulate; spire short; whorls 5, a little convex, the last inflated; aperture rounded-lunar, lip thickened within, sometimes subdentate on the columellar portion; umbilicus narrow, nearly covered by an expansion of the lip. White, varied by series of pellucid spots, sometimes running into stripes.

Diam. 16, height 13 mill.

Southern California.

Dr. Newcomb states that this species is not Californian or Oregonian, but belongs to a more southern fauna. I have specimens, however, referable to this species, received from Dr. Newcomb, from Bay of Monterey, Cal., as a variety of *H. areolata*. It is more orbicular than that species, the columella more distinctly tuberculate, and the surface more granulate. The first figure is a copy of that given by Pfeiffer, while the last represents a fresh and larger specimen.

---

7. **Polymita varians**, Menke.
Plate 6, figures 9-13.

Globose-conic, solid, smooth, shining, delicately striate; spire elevated-conic; whorls $5\frac{1}{2}$, convex, the last broadly rounded; aperture small, two-thirds circular, lip expanded a little, thickened within, very nearly covering the umbilicus; base convex. White, greenish, reddish or brown, sometimes with black or white bands, one or more in number, disposed on different portions of the surface, apex and columellar part of the lip always rose-color.

Diam. 17, height 15 mill.

Florida. (From West Indies.)

TACHEA, Leach.
1. **Tachea hortensis**, Müller.
Plate 6, figures 14, 15.

Subglobose, smooth; spire conoidal; whorls 5, the last ventricose; aperture rounded-lunar, lip dilated, thickened within; base convex, imperforate. Yellowish, sometimes with one to five revolving dark brown bands.

Diam. 19, height 16 mill.

New England States near the sea, and Islands on the coast. (From Europe.)

POMATIA, Leach.
1. **Pomatia aspersa,** Müller.
Plate 6, figure 16.

Subglobose, rather thin, coarsely and irregularly striate, finely striate and finely wrinkled and indented; spire obtuse; whorls 4–5, moderately convex, rapidly increasing, the last very large and ventricose; aperture large, oblique, rounded-lunate, lip white, sharp, a little expanded, extremities connected by a thin callus; umbilicus covered; base very convex. Yellowish or brownish, with brown bands, crossed by narrow undulating flammules of yellow.

Diam. 31, height 25 mill.

At various places on the Coast, New Orleans, Charleston, Maine, Nova Scotia. (From Europe.)

MESODONTINÆ.

Genera.

* *Shell minute, margin of lip circular, its extremities approaching and connected by a callus: not toothed.*

1. VALLONIA, Risso. Shell minute, diaphanous, umbilicate; lip margin broadly reflected, nearly circular, white.

European; extending into the States east of the Rocky Mountains.

** *Shell large, umbilicate, lip with a small tooth at its base; sometimes also with a small parietal tooth.*

2. ULOSTOMA, Albers. Shell large, globosely depressed, aperture semicircular, lip tuberculately toothed at base. Horn color, sometimes banded with rufous.

Alleghany Mountains, from Vermont to Tennessee, and westward to the Rocky Mountains.

*** *Shell large, umbilicus covered or perforate, lip not toothed; with generally a small oblique parietal tooth.*

3. MESODON, Rafinesque. Shell large, subglobose or orbicularly depressed; aperture rounded lunar, the lip sometimes slightly dentately thickened at the base; parietal tooth, when present, small; umbilicus either covered by an expansion of the lip or partially covered. Generally uniform pale horn color.

Inhabits the entire United States, but two species only on the Pacific slope of the continent.

**** *Shell of moderate size, turbinate or depressed; umbilicus closed; aperture trigonal, lip with a long lamellar tooth at base, and frequently a small tubercular tooth above, a large curved lamellar parietal tooth.*

4. XOLOTREMA, Rafinesque. Shell uniform horn color, depressed or turbinate, frequently angulate or carinate on the periphery; base convex; aperture with always a lamellar curved parietal tooth, and a long lamellar basal tooth, with frequently also a small denticle on the superior part of the lip.

Inhabits from the Alleghany to the Rocky Mountains.

***** *Shell moderate in size, aperture trilobate, caused by denticles on the superior and inferior parts of the lip, and on the parietal wall.*

† *Umbilicus open.*

5. TRIODOPSIS, Rafinesque. Shell globosely depressed, umbilicus open, lip teeth small, nearly equidistant. Uniform horn color.

†† *Umbilicus closed.*

6. ISOGNOMOSTOMA, Fitzinger. Shell smaller, globosely depressed, umbilicus covered by the extremity of the lip; aperture three-lobed, the lip teeth small, the parietal tooth larger, blade-shaped. Horn color, frequently hirsute.

Eastern, Middle and Southern States, also Europe.

****** *Shell small, aperture narrowly transverse, basal, extending from the periphery to the axis of the shell; parietal wall with a long lamellar tooth, lip broad, with generally a notch in the centre.*

7. STENOTREMA, Rafinesque. Shell small, generally hirsute, horn color, depressed turbinate above, very convex below; aperture narrow and long, basal, lip and parietal wall subparallel, the former with a long blade-shaped tooth, the latter either similar, with generally a notch in the middle, or, the notch being wider, with two teeth; umbilicus closed. Within the aperture, and near the axis, may be seen an accessory column or pillar, probably designed to assist the animal in retiring within its shell.

United States, east of Rocky Mountains.

******* *Shell small, depressed and ribbed-striate above, many-whorled; periphery generally carinate; convex beneath, showing several whorls; aperture with a V-shaped parietal tooth.*

† *Aperture tridentate, base exhibiting one and a half to two whorls.*

8. DÆDALOCHILA, Beck. Shell small, depressed, ribbed-striate, periphery angulate, below convex, showing more than one, sometimes nearly two, whorls, with a minute central perforation; lip auricular, frequently expanded, the place of the teeth being marked externally by scrobiculations; parietal tooth V-shaped, joined by a raised callus with the extremities of the lip.

Southern States, Mexico, Guatemala, Cuba, &c.

‡ *Shell planorboid, many-whorled; aperture with a V-shaped tooth, but no lip teeth.*

9. POLYGRA, Say. Shell planorboid, many-whorled, whorls narrow, ribbed above, periphery angulate; aperture small, subtrigonal, with a V-shaped parietal tooth, joined by a raised callus to the extremities of the lip; below plane, showing several whorls, with a narrow umbilicus. Horn color.

Gulf States and West Indies, also South America.

VALLONIA, Risso.

1. Vallonia minuta, Say.

Plate 7, figure 2.

Spire depressed, convex, whorls four, the last rapidly increasing and spreading at the mouth; thin, transparent, very minutely striated, or sometimes distantly costate; aperture orbicular, lip large, well rounded; umbilicus large. Light horn color, with generally a greenish tinge, lip white.

Diam. 2·5 mill.

Maine to South Carolina, and westward to the Rocky Mountains.

This species was described many years ago by Mr. Say, and his species has since been recognized as distinct by several conchologists, but the weight of opinion has been in favor of considering it identical with *Vallonia pulchella* of Europe. Mr. E. S. Morse has recently critically compared the two species, and discovered several differences which we have been able to confirm fully. *V. minuta* is more depressed, the whorls are not as large, the aperture wider, and the labrum not so much rounded above, while below it ends further towards the axis of the shell; the lip of *V. minuta* is at an angle of 27° from a line passing through the axis, while that of *V. pulchella* is 35°. The lingual dentition also differs. In order to exhibit the above differences in the shell satisfactorily, I have copied Morse's figures of *V. minuta* and *V. pulchella*, fig. 1 representing the latter. European authors have separated from *V. pulchella* a species in which the striæ of growth are occasionally elevated into ridges, under the name of *V. costata*. I doubt the validity of this distinction, which is of importance to those American conchologists who maintain the identity of our shell with the European *pulchella* from the fact that in certain localities we also have the costate variety. *Helix alternata* and several other species of native Helices exhibit quite as great diversity as the *minuta* in this respect, and I am inclined to attribute the development of these ridges in the growth of the shell to local disturbing influences.*

ULOSTOMA, Albers.

† *Shell banded ; no parietal tooth.*

1. Ulostoma profunda, Say.

Plate 7, fig. 3.

Orbicularly depressed; whorls 5—6, convex, strongly obliquely striate, with well-impressed suture; aperture subcircular, lip large, white, its extremities approaching with an obtuse tooth on the inner basal edge; base convex, umbilicus large, profound. Light horn color, with generally a broad reddish brown band above the periphery, and numerous narrow bands on the base, sometimes uniform pale horn color.

Diam. 28 mill., height 15 mill.

Western New York to Virginia, and eastwards to Nebraska.

---

* "The molluscous fauna of Harper's Ferry is distinguished for the development of heavy lines of growth, and acute prominent carinæ on the shells of the species; and in the terrestrial shells by the depression of the spire." Tryon *on the Mollusca of Harper's Ferry, Va.*, Proceed. Acad. Nat. Sci. Phila., 1861. The same features obtain in the species of the mountainou district of East Tennessee.

** *Unicolored, with a parietal tooth.*

2. Ulostoma Sayi, Binney.

Plate 7, figure 4.

Orbicularly depressed, whorls 5—6, thin, regularly obliquely striate; aperture suborbicular, lip white, the margin narrowly reflected and tuberculately dentate on the base, the parietal wall toothed; umbilicus moderate and deep. Pale horn color.

Diam. 22, height 15 mill.

Maine to Illinois, and southwards to Pennsylvania, inhabiting mountains and elevated districts.

MESODON, Rafinesque.

This group embraces most of the larger species of Helices inhabiting North America east of the Rocky Mountains. In Cuba it is replaced by *Pachystoma*, a genus very closely allied in the form, size and coloration of the shell; (but in Jamaica the larger species of shells belong to *Pleurodonta*, quite a different type in every respect). A like alliance brings very closely together the species of *Ulostoma* just described, with the numerous European genus *Campylæa*, so that we have in the subfamily *Mesodontinæ* first an undoubted American representative of *Vallonia*, then a magnified repetition of the same form, with modifications in *Ulostoma* which is very close to European species. This type of shell in this country appears to have become further modified into the genus *Mesodon*, in which form it has flourished exceedingly. Upon tracing *Mesodon* southwards, we find the species becoming larger, heavier and more coarsely striate, and these changes culminate in *Pachystoma*. The curious relation of the Pachystomæ with extinct and expiring species of land shells of Madeira is another curious fact in conchological geography. I shall have occasion, more than once before leaving the genera composing the subfamily of which *Mesodon* is the type, to point out among the terrestrial shells of Europe stray individuals of undoubtedly American forms.

* *Dentate.*

† *Umbilicus covered.*

Mesodon albolabris, Say.

VARIETY DENTATA.

Plate 7, figure 6.

This species not unfrequently develops a tooth on the parietal wall, (see description under section **,) and in this state it closely resembles *M. exoleta*, but may be distinguished by its larger size, less convex body whorl, broader lip, more transverse aperture, and generally lighter substance.

The dentate variety of *albolabris* has been found in Maine, Massachusetts, Pennsylvania, Ohio, Iowa, Illinois, Michigan, &c.

1. Mesodon exoleta, Binney.

Plate 7, figure 8.

Ventricose, suborbicular, whorls 5—6, convex, texture heavy, suture well marked; body whorl large and convex; aperture rounded, lip reflected, the parietal wall toothed; umbilicus covered by the extremity of the lip. Light horn color.

Diam. 25, height 15 mill.

New York to Georgia, west of the Alleghany Mountains, and extending westward to Missouri.

2. Mesodon dentifera, Binney.

Plate 7, figure 9.

Depressed, spire flatly convex, base well rounded, whorls 5, delicately striate, suture distinct, but not deeply impressed; aperture wide, lip broadly reflected and covering the umbilicus, parietal wall armed with a prominent tooth. Yellowish horn color.

Diam. 19 mill.

Maine to Virginia, and westward to Ohio, inhabiting mountain ranges and highly elevated ground.

Readily distinguished from the former species by its more depressed form, &c.

3. Mesodon Wheatleyi, Bland.

Plate 7, figure 10.

Conoidally globose, spire somewhat elevated, with distinct sutures, thin, closely ribbed-striate, with microscopic granulations, hirsute; whorls $5\frac{1}{2}$, convex, the last well rounded, but slightly depressed at the aperture, and constricted; aperture obliquely lunate, with the parietal wall armed by a tooth; base convex, umbilical region excavated, but imperforate. Reddish horn-colored, the lip rose-colored.

Diam. 14, height 7 mill.

Mountains in Cherokee County, North Carolina.

Mr. Bland remarks upon the close relationship existing between this species and *M. Columbiana* which inhabits the Pacific States. Each is the only hirsute species of *Mesodon* of its region. The two are also allied in size, form and color.

---

4. Mesodon Christyi, Bland.

Plate 7, figure 11.

Depressed, spire obtuse, whorls $4\frac{1}{2}$, somewhat convex, descending at the aperture, texture solid, with close, oblique, rib-like striae, periphery a little angular; aperture depressed, with a strong parietal tooth; base convex, excavated in the middle, imperforate. Dark horn color.

Diam. 10, height $4\frac{1}{2}$ mill.

Mountains in Cherokee County, North Carolina.

Mesodon Roemeri, Pfeiffer.

This species is very rarely furnished with a completely-covered umbilicus. See description in the perforate section, species 8.

†† *Umbilicus open.*

‡ *Shell rounded.*

5. Mesodon thyroides, Say.

Plate 8, figure 1.

Rounded, thin, regularly and closely obliquely striate; spire depressed, conical, suture distinct, but not deeply impressed; whorls 5, convex, the body large, well rounded, slightly declining at the aperture, behind which it is a little constricted; aperture obliquely semilunar, the lip moderately reflected, with a slight transverse tooth or varnish of callus near the top of the parietal wall; umbilicus open, but partially bounded by a raised acute dilatation of the lips. Uniform light horn or straw color.

Diam. 25, height 16 mill.

Inhabits plentifully the whole country east of the Rocky Mountains, but is particularly numerous in the Western States, becoming rare in New England, and partially replaced by *M. bucculenta* in the South.

This is one of our most beautiful species, its delicate texture, regular rib-like striae, pleasing color and frequently roseate lip, give it a particularly neat aspect. Until it has become quite mature, the only trace of the parietal tooth is frequently a slight uncolored thickening, appearing as though varnished. In the immature state it is likely to be confounded with young individuals of a small variety of *M. albolabris*, a species which always inhabits with it. It has also been very generally confounded with *M. bucculenta*, a smaller, more globose shell, the description of which follows.

---

6. Mesodon bucculenta, Gould.

Plate 8, figure 2.

Globose, rather thin, with delicate, oblique, regular raised striae; spire obtusely elevated, suture distinctly, but not deeply, marked; whorls 5, convex, body large, rounded compactly, a very little deflected towards the aperture; mouth lunate, with a broad white lip, partially covering the umbilicus; parietal tooth sometimes scarcely developed, but occasionally strong. Reddish horn color.

Diam. 16, height 10 mill.

Georgia, to Texas, Maryland, Eastern Pennsylvania, Western New York.

This species is allied closely to *M. thyroides*: it differs in its smaller size, more globular form, darker color and less open umbilicus. It was first described as a far Southern species, but has been recently ascertained to be common in the Middle States. A remarkable minor form of *thyroides* was described by Mr. W. G. Binney several years ago, from the vicinity of Philadelphia, and he declares his inability to distinguish it from forms of *bucculenta*. This shell is really *bucculenta*, and its identity was first pointed out by Dr. E. Michener, in the American Journal of Conchology, 1865. It has also been noticed in New York by DeKay, who described it in his Report under the name of *Helix rufa*. In the neighborhood of Philadelphia all the specimens of so-called *thyroides* that I have seen are the true *bucculenta*.

7. Mesodon devia, Gould.

Plate 8, figure 3.

Orbicularly depressed, rather thick, strongly obliquely striate; spire convex, suture moderately impressed; whorls 6, the last well rounded; aperture obliquely and transversely lunate, with a widely-reflexed white lip, which is somewhat toothed at the base, and impinges on the open umbilicus; parietal tooth trigonal, oblique, quite large. Dark horn color, nearly brown.

Diam. 20, height 11 mill.

Oregon.

Mesodon Columbiana.

VARIETY DENTATA.

Plate 8, figure 12.

This, which will be described among the non-dentate species, sometimes developes a parietal tooth. Mr. Bland, several years ago, (Annals N. Y. Lyceum, vii.,) mentioned that his cabinet contained a specimen with such a tooth. I have seen several specimens of this variety, and my cabinet contains them from three different localities.

‡‡ *Shell subangulate on the periphery.*

8. Mesodon Roemeri, Pfeiffer.

Plate 8, figure 4.

Shell depressed, rather thin, semi-transparent, closely, but faintly striate; spire a little elevated, suture slightly impressed: whorls 5, somewhat convex, the last one subcarinate or angulate on the periphery, scarcely descending to the aperture; aperture obliquely lunate, the lip well thickened, but hardly expanded above, though becoming towards the base well-reflected, covering partially the umbilicus, and rarely entirely closing it; parietal wall generally armed with a well-developed tooth. Horn-colored.

Diam. 21, height 10 mill.

Texas.

May be distinguished at once from all the other species by its depressed form and angulate periphery.

** *Not dentate.*
† *Umbilicus closed.*
‡ *Unicolored.*

9. Mesodon major, Binney.

Plate 8, figure 5.

Ventricose, convex, globosely turbinate, heavy, covered with coarse oblique striæ; spire, elevated, convex, suture well impressed; whorls 6, convex, the body whorl very large and subglobular, very slightly declining; aperture small, rounded lunate, lip thick, moderately wide, with a tooth-like elevation at the base near the body whorl, dilated and covering the umbilicus. Yellowish brown.

Diam. 44, height 33 mill.

Tennessee to Florida and Alabama.

Dr. Binney first described this as a species distinct from *M. albolabris*, and separated it on account of its larger size, more globose, elevated form, rounder aperture, coarser striæ, &c. Messrs. W. G. Binney, Newcomb, Gould and myself admit its validity, but Mr Bland, on the contrary, unites the two, believing that *major* is only *albolabris* living in situations highly favorable to its growth, and gives measurements of specimens from different northern and southern localities, showing a gradual increase in size, as well as variations of form. I believe that the true *major* inhabits the far Southern States, where it replaces *albolabris*, and I doubt very much whether it was ever found North of Southern Tennessee.

10. Mesodon albolabris, Say.

Plate 7, figures 5, 6, 7.

Depressed orbicular, moderately thick, closely obliquely striate, with crowded, slightly-impressed revolving lines; spire convex, suture not deeply impressed; whorls 5—6, flattened convex, the body a little deflected at the aperture, and contracted behind the lip; mouth lunar, with a widely-reflected white lip, which, at the base, covers the umbilicus. Yellowish brown or light chestnut color.

Variety with a parietal denticle.

Diam. 25, height 15 mill.

Inhabits from Canada to South Carolina, and westward to Arkansas and Nebraska.

One of our most common species. This, as well as all others of the group, when immature, is furnished with an open umbilicus and a sharp unreflected lip; and in this state it is exceedingly difficult to distinguish the species one from another.

11. Mesodon Pennsylvanica, Green.

Plate 8, figure 9.

Turbinately subglobose, moderately thick, translucent, with crowded, elevated oblique striæ; spire convexly elevated, suture distinct; whorls 6, convex; aperture subtriangular, contracted behind the lip, which is white, and narrowly reflected, and slightly thickened internally at the base; umbilical region indented, umbilicus covered. Bright horn color.

Diam. 13—18 mill.

Western Pennsylvania, Ohio, Illinois.

This species is distinguished from *M. clausa* by its imperforate base and triangular aperture, and from the following species by its more turbinate form, as well as by the aperture.

12. Mesodon Mitchelliana, Lea.

Plate 8, figure 10.

Subglobose, moderately thick, translucent, finely striate; spire convexly conical, suture not deeply marked; whorls 5, the last well rounded, contracted behind the lip above, but the groove becoming indistinct towards the base; aperture rounded, lip white, narrowly reflected; base very convex, imperforate. Light horn color.

Diam. 15, height 9 mill.

Found only in Ohio.

For the distinction between *M. Pennsylvanica* and this species, see the description of the former.

---

13. Mesodon divesta, Gould.

Plate 8, figure 11.

Depressed orbicular, moderately thick, coarsely obliquely striated; spire a little convex, with well-marked suture; whorls 6, the last subangulate at the periphery; aperture lunate, very oblique, lip white, broadly reflected, horizontal at base, its outer portion flexuous; base convex, umbilical region excavated, but covered. Dingy horn color.

Diam. 18, height 7 mill.

Washington Springs, Arkansas.

Distinguished by its depressed form. Although described nearly twenty years ago, no other locality than that above given has been discovered.

---

‡‡ *Banded.*

14. Mesodon multilineata, Say.

Plate 8, figure 8.

Depressed orbicular, rather thin, closely obliquely striate; spire convexly conical, suture deeply impressed; whorls 5—6, very convex, the last considerably deflected at the aperture; aperture lunate, the lip white, narrowly reflected and dilated into and covering the umbilicus. Horn color, with more or less numerous reddish brown revolving bands of varying width.

Diam. 25, height 16 mill.

Western and North-western States.

This is one of our most beautiful species, and is extremely numerous in the region inhabited by it. A variety is occasionally met with, having a uniform dark brown epidermis, (middle figure,) and Dr. Binney mentions having seen one or two specimens in which the epidermis was pure white.

†† *Umbilicus open.*

‡ *Color mottled.*

15. Mesodon Townsendiana, Lea.

Plate 8, figures 6—7.

Orbicular, rather thick, with oblique, irregular, coarse striæ and fine revolving lines, the body whorl malleated obliquely; spire convexly conical, suture moderately impressed; whorls $5\frac{1}{2}$, the last somewhat deflected at the aperture; aperture lunate, oblique, lip white, very much thickened; base convex, umbilical region subangulate, umbilicus moderate. Yellow and brown variegated by the malleations.

Diam. 30, height 18 mill.

Oregon.

A smaller, more compact, more elevated, not malleated shell, which, if not distinct, may be designated as *variety minor*; inhabits Idaho and Nebraska, (fig. 6).

‡‡ *Unicolored.*

§ *Hirsute.*

16. Mesodon Columbiana, Lea.

Plate 8, figures 12, 13, 14.

Depressed orbicular, covered with short close hairs arranged in lozenge; spire depressed conical, suture well impressed; whorls 6, the last rapidly increasing, very convex, deeply constricted behind the lip and descending to the aperture; aperture narrowly lunate, lip white, widely reflected, partly covering the umbilicus; base very convex. Yellowish horn color.

Diam. 16, height 10 mill.

Oregon.

Peculiar in this group, for its very close arrangement of short bristles all over the surface; the old shells are frequently denuded of these, but then the scars of their insertion are distinctly visible with a glass. As already mentioned, this species sometimes developes a strong, oblique, parietal tooth. (fig. 12).

---

§§ *Not hirsute.*

17 Mesodon Downieana, Bland.

Plate 8, figure 15.

Subglobose, thin, subpellucid, with obsolete rib-like striae and crowded minute revolving lines; spire depressed conical, suture moderately impressed; whorls 5, convex, the last tumid, scarcely descending, constricted behind the lip; aperture obliquely lunate, lip white, reflected narrowly, nearly covering the umbilicus; base convex. Greenish horn color.

Diam. 10·5, height 6 mill.

University Place, Franklin County, Tennessee.

Like *M. Christyi*, Bland, in form, but has no parietal tooth.

---

18. Mesodon clausa, Say.

Plate 8, figure 16.

Subglobose, moderately thick, with fine oblique striae; spire convex, with distinct suture; whorls 5, convex, the last large, contracted behind the lip; aperture rounded, lip white, narrowly reflected and nearly covering the umbilicus; base very convex. Light yellowish brown, shining.

Diam. 13—15 mill.

Inhabits all the Western States from the Great Lakes to Mississippi and Alabama.

This species is of the same size as *M. Pennsylvanica* and *M. Mitchelliana*, inhabiting partially the same region. Its distinctive characters have been already pointed out. It is most abundant in the southwest in a semi-fossil condition.

XOLOTREMA, Rafinesque.

There are but five species known to belong to this section of the Helices; characterized by a lamellar tooth on the base of the aperture, and a covered umbilicus. We first indicated its generic value, and gave a list of species belonging to it, in the American Journal of Conchology, p. 81, 1865. These shells inhabit the middle region of the United States rather sparingly, being nowhere very numerous. There are two groups of species, two of them being somewhat rounded trochiform, while the other three are depressed, and generally furnished with an additional tooth on the upper part of the labrum, making the aperture tridentate.

Shell elevated.

1. **Xolotrema elevata,** Say.

Plate 9, figure 1

Shell convexly conical, thick, finely obliquely striated; spire elevated, with a well-impressed suture; whorls nearly seven, convex, slowly increasing, the body large and well rounded; aperture somewhat triangular, contracted by the lip, which is thickened, but not very broadly reflected, and covers the umbilicus; a long lamellar tooth occupies nearly the whole basal part of the lip, and the pillar lip is furnished with a stout tooth, curving inwards above; the extremities of the labrum frequently connected by a well-defined callus. Yellowish horn color.

Diam. 23, height 16 mill.

From Western New York to West Virginia, and westward to Missouri.

Mr. W. G. Binney received two specimens, collected by the late Major Kennicott in Wisconsin, which were remarkable from having each a brown band revolving upon its periphery.

2. **Xolotrema Clarkii,** Lea.

Plate 9, figure 2.

Shell globosely conical, moderately thick, finely striated; spire obtusely conical, suture moderate; whorls seven, convex, slowly increasing, the body well rounded; aperture subtriangular, the lip reflected and thickened, entirely covering the umbilicus: a long lamellar tooth on the base of the lip, and a strong curved tooth on the parietal wall. Reddish horn color.

Diam. 14, height 9 mill.

Cherokee County, North Carolina.

A beautiful little species, almost a pigmy *elevata;* the whorls of the spire are not so convex, however, as they are in that species.

** *Shell depressed.*

3. Xolotrema obstricta, Say.

Plate 9, figure 3.

Shell depressed, acutely carinated, the carinæ visible on all save the apical whorls of the spire; spire depressed convex; whorls five, more convex below, covered with distant sharp oblique costæ, which fringe the edge of the carina in crossing it, with frequently slight, close, revolving lines; aperture trilobate, lip widely reflected, parietal tooth strong, oblique, superior lip tooth small, inferior lip tooth a long blade upon the base of the lip; umbilicus covered. Pale to dark brown.

Diam. 22, height 8 mill.

Western and Southern States.

This species differs from the following in being *always* strongly carinate, and in not having hispid epidermal projections.

---

4. Xolotrema palliata, Say.

Plate 9, figure 4.

Shell depressed above, a little more convex below, with elevated oblique ribs, the epidermis rugose, with close, minute hispid prominences; whorls five, not very convex, subangulate to carinate at the periphery; aperture trilobate, caused by three teeth, the largest of which is parietal and oblique, the others are respectively on the upper and lower portions of the labrum, the lower one being blade-shaped; umbilicus covered. Light to dark brown.

Diam. 23, height 10 mill.

Alleghany Mountains and westward, Vermont, (W. G. Binney,) Iowa, Michigan, Kentucky, Tennessee, &c.

5. Xolotrema appressa, Say.

Plate 9, figures 11, 7.

Shell depressed; spire only slightly raised, suture distinct; whorls 5, a little convex, the last subangulate at the periphery, and constricted behind the widely-reflected lip of the aperture; obliquely ribbed-striate above, smooth beneath; aperture trilobate, with a long oblique parietal tooth and two small lip teeth, of which one is on the upper and the other on the lower part of the lip; umbilicus filled by a deposit of callus. Yellowish horn color.

Diam. 18 mill.

Principally west of the Alleghany Mountains, from New York to Georgia, and westwards to Alabama, Tennessee and Illinois. Wilmington, N. C., and City Point, Va., (Thomas Bland).

Much smaller in size, and without the hispid surface of X. *palliata*; it is also flatter. Sometimes the lip teeth are not well developed. (Fig. 7.)

TRIODOPSIS Rafinesque.

This genus differs from the foregoing in the following particulars:—The umbilicus is open, the basal tooth of the tridentate aperture is tubercular instead of lamelliform, and the shell is smaller.

The species are widely distributed, some of them being found in all parts of the United States, and a few as far to the southwards as Central America. The species of *Triodopsis*, like those of some other groups, in their distribution seem to mutually replace one another, thus forming, in the opinion of the older writers, geographical varieties rather than species. That several of them had a common parentage is obvious; but, inasmuch as the departure of each from the common type, though slight, appears to be permanent, these forms must be regarded as true species.

1. Triodopsis tridentata, Say.

Plate 9, figures 6, 13.

Shell depressed, spire somewhat convex, suture moderately impressed; whorls 5—6, a little convex, obliquely ridged-striate; aperture trilobate, lip widely reflected, on the parietal wall is placed an oblique tooth, and small marginal teeth are on the upper and the basal portions of the labrum respectively; base convex, umbilicus open and deep. Horn color.

Diam. (large var.) 23, height 9 mill.; (small var.,) diam. 13, height 5·5 mill.

Inhabits all the States east of Rocky Mountains.

This shell is generally larger than the next, from which it differs in having a more depressed form and much smaller teeth.

---

2. Triodopsis fallax, Say.

Plate 9, figure 12.

Shell convex, spire somewhat elevated, suture distinct; whorls 5, moderately convex, obliquely coarsely striate; aperture trilobate, the parietal lamelliform tooth large and oblique, the lip teeth large and flattened, the upper one deflected into the aperture; base moderately convex, umbilicus open. Light horn color.

Diam. 11, height 6 mill.

Inhabits the whole country east of Rocky Mountains.

This shell, as mentioned under the description of *T. tridentata*, differs from that species by its smaller size, more elevated spire and larger teeth; the upper lip tooth is remarkable, being quite large, somewhat blade-shaped, and its surface bent into the aperture. I regard this shell as *fallax*, in deference to the opinions of Messrs. W. G. Binney, Bland, &c., but the fact is that Say confounded two species under this name, and his *description* is actually drawn up from a specimen of *introferens*, var. *minor*, as *fallax* is not found in the vicinity of Philadelphia—the habitat given by Say.

---

3. Triodopsis introferens, Bland.

Plate 9, figure 5.

Depressed, thin, with rib-like striæ; spire convex, but slightly elevated; whorls 5—6, somewhat rounded, the last descending, constricted behind the lip, with exterior pits marking the position of the lip teeth within the aperture, periphery subangular; aperture trilobate, parietal tooth oblique, lamelliform, basal lip tooth submarginal, lamelliform, continued within the aperture, where it forms a strong white tubercle, upper lip tooth transverse, and opposite to the external periphery; base convex, umbilicate, grooved within the umbilicus. Yellowish horn color.

Diam. 11—15, height 6—7 mill.

North Carolina; vicinity of Philadelphia, Pa., (small variety). Distinguished immediately from *tridentata* and *fallax* by its narrower umbilicus and groove, and especially by its tubercular basal tooth. See description of *T. fallax*.

4. Triodopsis Hopetonensis, Shuttleworth.

Plate 9, figure 9.

Depressed, thin, ribbed-striate; spire obtusely convex, but not much elevated, suture distinct; whorls 5½, somewhat convex, more convex beneath, constricted behind the aperture; aperture trilobate, a sharp parietal tooth and a small submarginal one on the upper and the lower lip; narrowly umbilicated. Light horn color.

Diam. 13, height 6 mill. (Var. major.)

Sullivan's Island, S. C.; Hopeton and St Simon's Isle, Geo.; Florida.

Distinguished from *fallax* by its smaller umbilicus, less thickened lip and teeth, and by the latter being more remote, one from another. The dimensions given above are those of a large specimen; ordinarily they attain but two-thirds of the size.

5. Triodopsis Yucantanea, Morelet.

Plate 9, figure 17.

Shell depressed, almost flat above, but quite convex below the angular periphery; whorls five, obliquely ridged above and striate beneath, constricted behind the reflected lip; aperture trilobate, parietal tooth very oblique, nearly V-shaped, with one lip tooth opposite to it, and a smaller erect one upon the centre of the basal portion of the lip; umbilicus rather wide and deep. Light horn color.

Diam. 8, height 3 mill.

Isle of Carmen, Yucatan.

This species is introduced to show that a form very closely allied to our own shells, inhabits a far-distant locality. Probably the intermediate country of Mexico contains undiscovered species exhibiting the same characteristics. The spire is represented too much elevated in the figure.

6. Triodopsis vultuosa, Gould.

Plate 9, figure 14.

Shell globosely depressed, closely delicately striate, rather solid; spire convex, not much elevated, suture well impressed; whorls 5½, convex, slowly increasing, the last subangular, and deflected at the aperture, well rounded below, with a moderate umbilicus; aperture lunate, its outline somewhat sinuous, parietal tooth a broad lamina, oblique, joined to the lip below, lower lip tooth small, upper lip tooth expanded and reflexed. Dark horn color.

Diam. 10, height 5 mill.

Texas.

Intermediate between *H. fallax* and *H. Texasiana*.

7. Triodopsis Mullani, Bland and Cooper.

Plate 9, figure 15.

Shell globosely depressed, irregularly striate, shining, with a thin epidermis, covered with minute spiral lines and tubercles, (the latter probably the scars of hairs); whorls 6, convex, much constricted behind the aperture, and smoother on the base; aperture trilobate, parietal tooth small, linguiform, lower lip tooth lamelliform, upper one small, and sometimes obsolete; umbilicus moderate, partially covered by the lip. Dark horn color.

Diam. 13·5, height 7 mill.

Washington Territory and Oregon.

Darker in color, with smaller umbilicus and differently formed teeth from *tridentata*.

8. Triodopsis loricata, Gould.

Plate 9, figures 16, 19.

Shell small, orbicularly depressed; spire convex, not much elevated, suture well impressed; whorls 5, convex, thin, obliquely striate, with small epidermal scales or scars, and in fresh specimens hispid, very much contracted behind the lip; aperture transversely trilobate, the very oblique parietal tooth quite small, and the two lip teeth merely slight elevations of the surface; base very convex, umbilicus narrow and deep, slightly circumscribed by the lip. Dark horn color.

Diam. 6, height 4 mill.
" 8, " 5 " (var. major.

California.

In general appearance this shell is singularly allied to *Stenotrema monodon;* it is distinguished principally by the small, scarcely-developed lip teeth, and by its geographical distribution. The roughened appearance of the epidermis, as it generally exists when denuded of hairs, being the scars of their insertion, is another link connecting this with *Stenotrema*.

ISOGNOMOSTOMA, Fitzinger.

This genus has for its type *Helix personata* of Lamarck, a common European shell, with which the two following species are closely allied. The shells are more elevated, and revolve more closely than in *Triodopsis*, appearing, when viewed from above, to be very like *Stenotrema*. The umbilicus is covered in this, differing from the last genus, as well as by the generally smaller size.

1. Isognomostoma inflecta, Say.

Plate 9, figure 10.

Shell depressed, convex; spire slightly raised, suture not deep; whorls 5, minutely obliquely striate, sometimes hirsute, very much contracted behind the lip; aperture trilobed, parietal tooth almost transverse, large and blade-shaped, basal lip tooth a small upright tubercle, upper lip tooth in the middle of the outer lip, a somewhat larger tubercle, and inflected, lip covering the umbilicus. Light horn color.

Diam. 11, height 4·5 mill.

Western Pennsylvania to Michigan, and southwards to Texas.

The constriction of the whorl behind the aperture is so great in this species that the reflected lip does not project beyond the general circumference of the shell.

2. Isognomostoma Rugeli, Shu'tleworth.
Plate 9, figure 8.

Shell convex, depressed, rather smooth; spire convex, but little elevated, suture well marked; whorls 5½, narrow, closely revolving, the last very much contracted in the centre behind the lip; aperture small, with a prominent bent, oblique parietal, lamellar tooth, a small upright tubercular basal tooth, and a large lamellar tooth opposite to the parietal tooth, and situated entirely farther within the aperture; base convex, umbilicus covered. Light horn color.

Diam. 10—13, height 5—6, mill.

Tennessee and North Carolina.

Easily distinguished from *I. inflecta* by its upper lip tooth, situated far within the aperture.

STENOTREMA, Rafinesque.

In this genus the shell is orbicular and generally hirsute, the whorls revolve closely, and the aperture is basal and narrowly transverse, extending from the periphery to the axis. In most of the species the dentition of the aperture in *Stenotrema* is peculiar to this genus—namely, a long transverse parietal blade and a parallel thickening of the basal portion of the lip, which is frequently incised in the middle.

Stenotrema inhabits the entire extent of the United States, and several of its species are widely diffused, whilst others, on the contrary, are extremely local, and two or three of them very rare and highly esteemed.

* *Umbilicus open, or partly covered.*

1. Stenotrema monodon, Rackett.
Plate 9, figures 18, 20.

Shell convex, depressed; spire slightly elevated, suture very distinct; whorls 5, convex, narrow, finely striated and minutely hirsute, or covered with the scars of the hairs, deeply grooved behind the reflected lip; aperture transverse, with a long oblique parietal tooth, outer lip narrowly reflected, its basal termination more or less encroaching on the umbilicus; under surface very convex, much impressed around the deep, narrow, more or less closed axis. Dark horn color.

Diam. 7—10, height $3\frac{1}{2}$—$5\frac{1}{2}$ mill.

Inhabits the whole country east of the Rocky Mountains.

This is a somewhat variable species, and varieties of it have been considered distinct by several American conchologists. The typical *monodon* is supposed to be restricted to those species of large growth and open umbilicus, while *fraterna*, of Say, is the name applied to those having the umbilicus covered. A more convex variety, with narrower whorls, and generally smaller size, is called *H. Leaii*, Ward. We are told that the latter affects moist situations, while the true *monodon* inhabits dry places. I agree with Messrs. Binney and Bland in believing that we have not yet sufficient data to justify the separation into species of these varieties of *monodon*.

** *Umbilicus closed.*

† *Periphery rounded.*

‡ *Outer lip incised in the middle.*

2. Stenotrema stenotrema, Ferussac.
Plate 9, figures 21, 30.

Shell subglobose, depressed; spire convex, somewhat conical, suture well impressed; whorls 5, well rounded, narrow, slowly increasing in size, subangulate on the periphery, more convex below, and slightly impressed in the umbilical region, finely striate, and covered with close short hairs; aperture very narrow, extending to the axis of the shell below, and almost closed by the long lamelliform, outwardly projecting parietal tooth, the narrow depressed outer lip is reflected close upon the whorl, with a small triangular notch in its centre. Chestnut brown, lips white or pink.

Diam. 10, height 6 mill.

Western and Southern States.

3. Stenotrema hirsuta, Say.

Plate 9, figure 24.

Shell subglobose, hairy; spire convex, elevated, suture deep: whorls 5, well rounded, periphery subangulate, shell very convex below, umbilicus covered; aperture narrowly transverse, nearly closed by the lamelliform parietal tooth, the outer lip with a triangular notch upon its basal portion.

Diam. 6, height 4 mill.

New England, Middle and Western States.

The following are the chief differences between this species and the preceding:—*Hirsuta* is smaller, more globose, its parietal tooth somewhat sinuous, and terminating abruptly, and the lip notch larger proportionally. *S. stenotrema* has a smaller and more central lip notch, and its large parietal tooth is regularly bow-shaped over its edge, instead of being sinuous and abruptly terminated. The species are both of them widely distributed, but the range of *hirsuta* is far greater than that of *stenotrema*.

‡‡ *Outer lip not incised in the middle.*

4. Stenotrema maxillata, Gould.

Plate 9, figures 31, 35.

Shell small, globose; spire conical, convex, suture well impressed, subangular on the periphery, and more convex below; whorls 5, narrow; aperture transverse, nearly filled by a long lamellar parietal tooth, lip closely appressed, narrow, with a lamina behind its margin, and scarcely visible on account of the parietal tooth being in front of it; this lamina tapers out to the margin of the lip at its superior termination; umbilicus covered. Light chestnut color.

Diam. 6, height 4 mill.

Tennessee, Chattahoochee River, Georgia.

Readily distinguished from *hirsuta* by its entire lip and the raised lamellar tooth behind it.

5. Stenotrema germana, Gould.

Plate 9, figures 22, 23.

Shell small, solid, imperforate; spire depressed conical above; whorls 5, narrow, subangular at the periphery, and very convex below; aperture narrowly transverse, the parietal wall with a long blade-shaped tooth. Horn color, with a few scattered hairs.

Diam. 7·5, height 5 mill.

Oregon.

Very like *S. monodon*, but the base is more convex, and not indented around the axis. The hairs are much fewer in number than in *S. hirsuta*.

---

†† *Periphery carinate.*

‡ *Outer lip incised in the middle.*

§ *Lenticular species.*

6. Stenotrema spinosa, Lea.

Plate 9, figures 26, 28, 29.

Lenticular, upper surface depressed conical, suture slightly marked; whorls 6, flat above, carinate at the periphery, and convex below, slowly increasing in size, and covered with prostrate hairs in fresh specimens; aperture very narrow, lip slightly reflected and thickened, slightly incised in the middle, parietal tooth long, narrow, projecting, extending from the axis to the angle of the lip above; umbilical region slightly indented. Dark chestnut color.

Diam. 14, height 5 mill.

Mountainous regions of East Tennessee, and the northern parts of Alabama and Georgia.

The revolution of the whorls of the spire causes a very slight projection of the carina of each at the suture. Young shells are widely umbilicate, with hairs covering the surface, and projecting around the periphery like a fringe.

7. Stenotrema Edgariana, Lea.
Plate 9, figure 27.

Shell somewhat lenticular; spire depressed trochiform, suture distinct; whorls 5, flattened above, periphery carinate, the base convex, imperforate; aperture narrowly transverse, the outer lip notched in the middle, the parietal lip with a long blade-shaped tooth. Dark brown, hairy when fresh.

Diam. 10, alt. 5 mill.

Cumberland Mountains, Tennessee.

Smaller, more elevated, and more convex beneath, than *S. spinosa*. The parietal tooth most resembles that of *S. stenotrema*, while the form of that of *S. spinosa* is more like that of *hirsuta*. Another difference is in the suture, which, in the present species, is well marked.

---

§§ *Elevated Species.*

8. Stenotrema Edwardsii, Bland.
Plate 9, figure 34.

Lenticular, imperforate, carinate, obsolete near the aperture, rather thin; spire slightly convex; whorls 5, narrow, slowly increasing, with flat or erect bristles on the epidermis, or their scars when denuded of them; base very convex, but slightly indented around the axis, with impressed spiral lines under the epidermis; aperture narrow, transverse, with a narrow, slightly-curved, blade-shaped parietal tooth, upper margin of the lip scarcely reflected, basal portion reflected a little, and appressed partially to the body whorl, with a tooth-like callus within, and an almost obsolete central notch. Dark brown.

Diam. 9, alt. 5 mill.

Mountains in Fayette, or Green Brier County, Virginia.

This shell differs from *S. hirsuta* in its angulated periphery, and less distinct notch in the lip.

---

9. Stenotrema labrosa, Bland.
Plate 9, figure 25.

Shell imperforate, lenticular, carinated, solid, finely obliquely striate, epidermis thin, with prostrate hairs when fresh; spire slightly raised, suture a little impressed; whorls 5½, narrow, slowly enlarging, the last deflexed and constricted behind the lip; aperture transverse, narrow, ear-shaped, the extremities of the lip connected by a callus and a long blade-shaped tooth upon the body whorl, lip with a deep, wide notch in the middle. Dark brown.

Diam. 12·5, alt. 6·5 mill.

Arkansas, Alabama and Tennessee.

Distinguished from *S. Edgariana* by the thickened and reflected lip, and its deep, wide notch.

‡‡ *Outer lip not incised in the middle.*

10. Stenotrema barbigera, Redfield.

Plate 9, figures 32, 33.

Shell somewhat lenticular, sharply carinate, the carinæ of the spire whorls overlapping the suture; spire convexly conical; whorls flattened, narrow, 5½ in number; base convex; aperture narrow, transverse, extending from the periphery to the axis, which is covered by the lip; lip reflected, not dentate or incised, parietal tooth, a long lamina running parallel with it. Dark horn color, epidermis striate, hirsute, forming cilia on the periphery.

Diam. 10, height 6 mill.

North West Georgia.

Like the other carinate *Stenotremæ*, this is a mountain species, and inhabits the same region. It is readily distinguished from its allies by the absence of a lip notch.

DÆDALOCHILA, Beck.

This is a very peculiar group, embracing quite a number of species of subtropical distribution, most of the species occurring in the southern Gulf States, and in Texas and Mexico; they pass the Rocky Mountains, and appear on the west coast of Mexico. Owing to the comparatively unexplored countries which they inhabit, the species of *Dædalochila* have generally been only recently described, and it is not at all unlikely that many additional ones remain to be characterized.

In general appearance the species are very closely allied, one with another, and the differences are generally found in the teeth and shape of the apertures.

* *Margin of aperture regular.*
† *With two small marginal teeth on its outer lip.*
‡ *Diameter = 6 millimetres.*

1. Dædalochila leporina, Gould.

Plate 10, figures 1, 4.

Shell small, lenticular, slightly hairy, minutely striate; spire convex, depressed; whorls 5, somewhat convex, the last subangulate at the periphery; base convex, the umbilicus nearly covered, umbilical region excavated; aperture lunate, lip incumbent, reflexed, with a central sinus, the sides of which are formed by two teeth, parietal tooth V-shaped. Chestnut color, the lip sometimes roseate.

Diam. 5, height 3 mill.

Georgia, Arkansas, Mississippi, Indiana, Illinois, Tennessee.

It will be seen that this species has a wide distribution, although it has only been noticed hitherto at one or two places in each of the States mentioned.

2. Dædalochila pustuloides, Bland.

Plate 10, figures 2, 3.

Shell small, planorboid, thin, delicately striate, slightly hirsute; spire scarcely elevated; whorls $4\frac{1}{2}$, narrow, a little convex above, subangular at the periphery, and quite convex below, gibbous, constricted and suddenly deflexed at the aperture; aperture lunate with a lamelliform parietal tooth joined to the upper extremity of the lip by a sharp callus, outer lip reflected, thickened within, with two internal teeth, and a deep notch between them; umbilicus wide and deep. Horn color.

Diam. 5·5, height 2·5 mill.

Alabama; near Darien, Georgia.

Respecting this species, Mr. Bland says, "*H. pustuloides* is intermediate in size between *H. pustula* and *H. leporina*, is less globose than the former, and more sparingly hirsute. It differs widely from both in the character of the umbilicus; the aperture is much like that of *pustula*, but more narrow than that of *leporina*. The inferior tooth on the peristome is more developed laterally than in *H. pustula*—indeed, it has a somewhat bifid appearance, in which respect it is more allied to *H. leporina*.

3. Dædalochila pustula, Ferussac.

Plate 10, figures 6, 17.

Shell small, depressed, lightly striate; spire scarcely raised, suture well impressed; whorls 4, convex, subargulate on the periphery, and more convex below it, deflected at the aperture; aperture narrow, arcuate, impinging on the umbilicus; teeth as usual in the group, viz., one V-shaped parietal tooth and two tubercular lip teeth margining a central notch; umbilicus small and deep, with a groove revolving within it, which forms a raised lamina on the basal part of the interior of the shell. Reddish brown, with short hairs.

Diam. 5, height 3 mill.

Texas, Georgia; St. Augustine, Florida.

This shell and *pustuloides* were confounded by authors until recently distinguished, and the latter described by Bland. *D. pustula* is distinguished by its narrower umbilicus, and the umlical groove, and corresponding raised internal lamina; the latter is only to be seen by breaking the whorl.

4. Dædalochila Texasiana, Moricand.

Plate 10, figures 5, 36, 38.

Shell orbicular, depressed, rather solid, ridged above, smooth below; spire scarcely elevated, suture moderately impressed; whorls 5, slightly convex above, subangular at periphery, quite convex below, deflected at the aperture; aperture crescentic, with the lip margins joined by a large V-shaped tooth on the body whorl, outer lip with two denticles closely placed, and a pit between them; umbilicus minutely perforate. Pale horn color.

Diam. 9, height 4 mill.

Texas; Tamaulipas, Mexico.

5. Dædalochila triodontoides, Bland.

Plate 10, figures 10, 31.

Perforate, orbicular, depressed, thin, subpellucid, obsoletely striate above, smooth beneath; whorls 5, somewhat convex, deflexed at the aperture, subangulate on the periphery, and very convex below; aperture lunate, oblique, the extremities of the reflected lip connected on the body whorl by a V-shaped tooth, the lip having two small teeth situated far apart, one on the circumference the other on the base. Pale horn color.

Diam. 9·5, alt. 5 mill.

De Witt County, Texas; Corpus Christi, Texas.

More delicate, not as distinctly ribbed, more elevated than *D. Texasiana*; the lip teeth also are smaller and farther apart.

6. Dædalochila ventrosula, Pfeiffer.

Plate 10, figures 39, 35.

Orbicular, depressed, minutely perforated, thin, shining; spire slightly raised, suture well impressed; whorls 5, but little convex, finely striate above, smooth and very convex beneath, the periphery subangulate, much constricted behind the aperture; aperture rounded lunate, the terminations of the lip joined by a V-shaped parietal tooth; basal portion of lip with two white denticles, and circumference with a large lamellar tooth. Horn color.

Diam. 8, height 4·5 mill.

Texas and Mexico.

Sometimes attains to one-half larger size.

7. Dædalochila Hindsi, Pfeiffer.

Plate 10, figures 44; 24.

Shell depressed, orbicular, narrowly umbilicate, finely striate, diaphanous, shining; spire somewhat conical; whorls 5, but slightly convex, more convex below, deflected at the aperture and constricted behind it; apertur lunate, the lip slightly reflected, and its extremities united on the body whorl by a callus and a V-shaped tooth; on the lip are two small basal teeth and a large lamellar tooth opposite to the parietal one. Color light corneous.

Diam. 8, alt. 4.5 mill.

Texas and Mexico.

‡‡ *Diameter* = 10 *millimetres.*

8. Dædalochila tholus, W. G. Binney.

Plate 10, figures 7, 9.

Shell solid, white, shining, ribbed above, smooth below: spire depressed, conical, suture distinct; whorls 7, convex, slightly angular on the periphery; umbilicus broad and shallow, about half the diameter of the shell, showing $2\frac{1}{2}$ grooved whorls; aperture semi-oval, with a thickened, scarcely reflected lip bearing one median and one basal tooth—both small, parietal tooth large, rhomboidal.

Diam. 11, height 4 mill.

Texas.

9. Dædalochila Mooreana, W. G. Binney.

Plate 10, figure 8.

Depressed, carinated, white, strongly striate above and nearly smooth beneath; spire somewhat raised, suture deep; whorls 6, the last much deflexed at the aperture; umbilicus narrower than in *D. tholus*, exhibiting $1\frac{1}{2}$ whorls fully; aperture semi-oval, lip broad, heavy, but slightly reflected with two marginal teeth, of which one is basal, and the other sub-basal, parietal tooth large, rhomboidal.

Diam. 10, height 3.5 mill.

Texas.

It is questionable whether this and the forgoing are not varieties of the same species. The only difference is in the narrower umbilicus, broader lip and lower position of the lip-teeth in *Mooreana*.

Mexican Species.

10. Dædalochila Behrii, Gabb.

Plate 10, figures 40, 41, 43.

Shell planorboid, coarsely ridged above, striate below; whorls 5, quite convex, subangular at the periphery and very much deflected and constricted behind the aperture; umbilicus broad with a minute central perforation; aperture rounded, the lip expanding, its extremities approaching and connected by a large V-shaped tooth, there is also a large marginal basal tooth, and a smaller tubercular tooth on the middle of the lip. White, (bleached).

Diam. 15, height 4.5 mill.

Guaymas, Mexico.

---

† † *With three small marginal teeth on the outer lip.*

* *

Mexican Species.

11. Dædalochila acute-dentata, W. G. Binney.

Plate 10, figures 11, 13.

Shell planorboid, whitish, smooth, whorls 6, very convex, spire scarcely elevated, suture distinct; the body whorl oblique, inflated, deflected at the aperture and scrobiculate behind the lip; aperture small, ringent, lip circular, its extremities joined by a broad angular tooth on the body whorl; there are three lip teeth, one of them on the basal edge, perpendicular and short, the other two are further within and form slight elongated horizontal laminæ, causing pits on the outer surface of the whorl.

Diam. 14, alt. 4 mill.

Cinaloa, on Mazatlan River, Mexico.

---

12. Dædalochila Loisa, W. G. Binney.

Plate 10, figures 12, 14.

Shell planorboid, whitish, scarcely striate, suture slight; whorls 5, the last somewhat swollen, deflected channelled, and with two pits behind the lips; aperture ringent, 5-dentate, one on the parietal wall, large subtrigonal, connecting the lip extremities, the heavy lip slightly reflected, is armed with four teeth, two of them short, stout and perpendicular on the edge near the columella, while superior to them and farther within are two short, slender horizontal laminæ.

Diam. 13, alt. 5 mill.

Cinaloa, on Mazatlan River, Mexico.

Perhaps only a variety of *D. acute-dentata*, in which an additional basal tooth takes the place of a scarcely tooth-like elevation in the typical *acute-dentata*.

13. Dædalochila Ariadne, Pfeiffer.

Plate 10, figures 15, 16, 18.

Shell depressed, subdiscoidal, finely ribbed above, smooth beneath, upper surface a little convex; whorls 5, with slightly depressed suture; body whorl deflected, much constricted and scrobiculate behind the aperture; aperture small, lip thick, much expanded, its extremities joined by a callus forming an irregular, flexuose, V-shaped tooth, its point extending far within the aperture; lip with two stout basal folds converging within, while behind them is a perpendicular broad laminæ almost entirely closing the mouth; base convex, showing more than one whorl with a rounded umbilical groove terminating in a minute oblique central perforation. Transparent white, shining.

Diam. 12, height 5 mill.

Tamaulipas, Mexico.

---

††† *Outer teeth tubercular, swelled, placed behind the parietal tooth.*

‡ *Shell, smooth below.*

14. Dædalochila Dorfeuilliana, Lea.

Plate 10, figures 20, 21.

Obtusely conical above and slightly angulated on the periphery; whorls 6, with well impressed suture, ribbed, smoother below, deflected and constricted behind the aperture but without scrobiculations. Aperture lunar, the extremities of the lip joined by a callus forming a quadrate tooth on the outer whorl, far within, behind the parietal tooth, are two rounded tubercles, one superior and one basal, of nearly equal size; base showing $1\frac{1}{2}$ whorls and a minute perforation. Whitish.

Diam. 7·5, height 4 mill.

Kentucky and Tennessee. Texas?

15. Dædalochila Jacksoni, Bland.
Plate 10, figures 32, 33, 34.

Depressed, striate, smoother beneath, narrowly umbilicate; whorls 8, slightly convex, deflected, and contracted behind the lip; aperture oblique, lunate, lip thick, briefly reflected, its extremities joined by a V-shaped parietal tooth, basal margin with a strong, oblique sinuous fold, right margin with a deeply seated tooth. Dark brown, shining, lip brownish.

Diam. 7, height 4 mill.

Fort Gibson, Indian Territory.

Fig. 32 represents an elevated variety.

---

16. Dædalochila fastigans. L. W. Say.
Plate 10, figures 22, 23, 26.

Convex beneath, nearly plane above; whorls 6, ribbed above, smooth beneath, periphery sharply carinated; aperture lunate the extremities of the margin connected by a V-shaped tooth, upper lip tooth compressed, transverse, situated remote from the margin, lower lip tooth compressed, marginal, the position of the two being marked by pits in the outer surface of the whorl. Brownish.

Diam. 8, height 3 mill.

Tennessee.

---

‡ ‡ *Shell striate below.*

17. Dædalochila Troostiana, Lea.
Plate 10, figures 19, 25.

Differs from *D. fastigans* by its less prominent carina, the rib-like striæ being well developed on the base and its parietal tooth more quadrangular. The spire is slightly, convexly elevated. Corneous when fresh, minutely hirsute.

Diam. 6, height 3 mill.

Tennessee.

18. Dædalochila Hazardi, Bland.
Plate 10, figures 27, 28, 29.

Depressed above, convex below; whorls 5, narrow, ribbed above and below, periphery not carinate, aperture sub-reniform; lip-extremities connected by a V-shaped parietal tooth, lip teeth deep-seated, the lower one the largest and partly obscuring the upper one, which is situated farther in; externally scrobiculate opposite to these teeth. Epidermis sparingly hirsute, brown.

Diam. 7, height 3 mill.

Kentucky, Tennessee, Georgia, Alabama.

Distinguished from *Troostiana* by not being carinate, and by the lower tooth, which runs blade-shaped into the aperture a short distance.

---

††† *Two parietal lamellar teeth running within the shell.*

Dædalochila? hippocrepis, Pfeiffer.
Plate 10, figure 42.

Shell heavy, depressed, opaque, with flattened spire and impressed suture; whorls $5\frac{1}{2}$, scarcely convex above; periphery angulate below, convex, abruptly reflected at the aperture and constricted behind the lip; umbilicus expanded and grooved, with a minute central perforation; aperture extending from the periphery to the umbilicus, somewhat ear-shaped, lip white, expanded, its extremities connected by a V-shaped tooth, the two laminæ of which run far within the shell; upper portion of the lip with an entering angle, basal portion callous and reflected.

Diam. 12, height 5 mill.

New Braunfels, Texas.

** *Aperture expanded, ear-shaped, its margin continuous; upper lip tooth hook-shaped.*

20. Dædalochila auriformis, Bland.
Plate 11, figures 1, 2, 3.

Depressed and ribbed-striate above, periphery subangular, convex and smooth beneath; whorls $5\frac{1}{2}$—6, suture moderate, the last one slightly deflected and expanded at the aperture into an ear-shaped lip, behind which it is constricted; the base shows nearly two whorls, and a minute perforation, and a deep groove revolving on the inner side of the whorls; aperture ear-shaped, the peristome continuous and expanded, parietal tooth linguiform, two lip teeth. the upper one of which is a submarginal perpendicular lamella, and the basal one an oblique fold. White or brownish horn color.

Diam. 10, alt. 5·5 mill.

Alabama; Texas.

Differs from *D. avara* in having an umbilical groove.

21. Dædalochila avara, Say.

Plate 11, figures 4, 5, 6.

Spire depressed, periphery slightly angular; whorls 5, slightly convex above, with moderate suture, coarsely ribbed, hirsute, convex below, and ribbed only for a short distance behind the lip, the balance of the base nearly smooth; base without groove, showing a little more than one whorl, and with a minute perforation; aperture with reflected lip, the extremities of which are joined by a callus, forming a large $\vee$-shaped tooth; right margin with two teeth, one of them basal and oblique, the other submarginal and lamellar.

Diam. 7, height 3 mill.

Florida.

22. Dædalochila espiloca, Ravenel.

Plate 11, figures 7, 8, 9.

Spire slightly elevated and a little convex, periphery subangular, well rounded below; whorls 5, thin, ribbed-striate above, finely striate below, suture moderate; the last whorl a little deflected and contracted behind the aperture; base showing $1\frac{1}{2}$ whorls, the umbilical region bounded by an angle; margin of aperture continuous, the parietal tooth linguiform, the upper lip tooth hook-shaped at its lower termination, the basal tooth oblique. Horn color, shortly hirsute.

Diam. 8, alt. 3·5 mill.

Sullivan's Island, South Carolina.

23. Dædalochila Postelliana, Bland.
Plate 11, figures 10, 11, 12.

Spire slightly elevated, conical; whorls 5, ribbed-striate above, finely striate below, periphery subangular; aperture ear-shaped, much contracted, the margins joined by a large linguiform parietal tooth entering the aperture, right margin with a deep-seated lamella terminating below in a hook, basal margin with a prominent oblique lamella extending to the edge of the lip. Brownish horn-color, thin, pellucid, aperture white.

Diam. 9·5, alt. 5 mill.

Georgia.

24. Dædalochila auriculata, Say.
Plate 11, figures 13, 14.

Spire low, conical; whorls 5, ribbed-striate above and below, the periphery slightly angulate; last whorl suddenly deflected to the aperture, and very strongly scrobiculate behind the middle of the right margin, and also at the base; aperture oblique, auriform, the parietal tooth large, irregular, projecting inwards and upwards, the right margin with a perpendicular lamellar tooth, and the base twisted into a large oblique tooth; base showing 1½ whorls, the umbilical region bounded by a carina.

Diam. 13, height 7·5 mill.

St. Augustine, Florida.

Larger than the other species, and distinguished by the ribs on the base.

25. Dædalochila uvulifera, Shuttleworth.
Plate 11, figures 15, 16.

Spire depressed conical; whorls 5, with close ribs extending entirely across the base; periphery subangulate, lower surface convex; last whorl deflected at the aperture and scrobiculate; aperture oblique, much contracted by teeth, with the lip very much expanded, parietal tooth quadrately linguiform, extending into the aperture, the upper lip tooth situated far within the aperture, the basal lip tooth an oblique plication; umbilical region subcarinate. Horn color.

Diam. 12, alt. 7 mill.

Florida; Corpus Christi, Texas.

Smaller, with differently formed aperture from *D. auriculata*.

POLYGYRA, Say.

This is a tropical genus, containing many flat, wheel-shaped species which inhabit the shores of the Gulf of Mexico, the West Indies and Mexico. The southern limit of the group extends to the vast regions of the empire of Brazil, whence come the curious *P. heligmoidea* and *P. helicycloidea*. Unlike *Dædalochila*, the aperture is simple and small, the lip reflected but not expanded, and not dentate, while the parietal tooth is small. The aspect of the shell is singularly like that of *Planorbis*, and reminds one strongly of several of the species of that fresh water genus inhabiting Europe. A thin, thread-like lamina occasionally revolves upon the inner wall of the aperture, and is visible through the whorl; it has been detected in most of the species, and probably is at times developed in all of them, though many specimens are without it.

1. Polygyra anilis, Gabb.
Plate 11, figures 17, 18.

Spire nearly flat, whorls $4\frac{1}{2}$, the last descending to the aperture, and a little constricted, suture well impressed, surface microscopically striate above and below; base showing about $1\frac{1}{2}$ whorls, with a minute central perforation. White.

Diam. 13, height 6 mill.

Guaymas, Mexico.

This is scarcely a typical *Polygyra*, as it does not exhibit so many whorls on the base as the other species. The aperture, however, is that of a *Polygyra*.

2. Polygyra cereolus, Mühlfelt.
Plate 11, figures 19, 20, 21.

Shell lenticular, the spire very nearly flat, under surface flat or slightly concave, periphery subangulate; whorls 7, flattened and closely ribbed above; base smoother, showing 5 whorls, with a narrow umbilicus; aperture small, subtriangular, the margins connected by a slight callus, developing in the middle into a small triangular tooth. Light horn color.

Diam. 14, alt. 3·5 mill., var. *major*.
" 9, " 2·5 " " *minor*.

East Florida.

Differs from *P. septemvolva* principally by its umbilicus being much narrower. (Compare fig. 19 with fig. 22.)

3. Polygyra septemvolva, Say.
Plate 11, figure 22.

Shell discoidal, with seven whorls, which are closely ribbed-striate above, the periphery subangular; smooth below, and exhibiting about four whorls besides those forming the walls of the rather large umbilicus. Horn color.

Diam. 14, height 3·5 mill.

Florida.

---

4. Polygyra Carpenteriana, Bland.
Plate 11, figures 23, 24.

Subdiscoidal, spire slightly elevated; whorls $5\frac{1}{2}$ to $6\frac{1}{2}$, thin, shining, closely obliquely ribbed above, smooth beneath the angular periphery; base showing $2\frac{1}{2}$ whorls, with a minute perforation. Light horn color or rufous, with frequently opaque irregular bands crossing the whorls.

Diam. 8—10, alt. 3—4 mill.

East Florida.

---

5. Polygyra volvoxis, Parreyss.
Plate 11, figure 25.

Spire a little elevated; whorls 7, flattened above, convex beneath, the periphery carinate; upper surface and base near the aperture closely ribbed, the balance of the base smooth; over two whorls visible below, besides a narrow umbilicus. Horn color, with frequently white blotches or bands running across the whorls.

Diam. 9, alt. 4 mill.

East Georgia and Florida.

This is perhaps a young state of *P. septemvolva*, Say.

6. Polygyra Febigerii, Bland.
Plate 10, figures 30, 33.

Depressed, spire scarcely raised; whorls $5\frac{1}{2}$—6, ribbed-striate above, finely striate below, periphery angulate; aperture subtriangular, with a small parietal tooth; base exhibiting about $1\frac{1}{2}$ whorls, with a central perforation. Pale reddish horn color.

Diam. 8·5, alt. 3·5 mill.

New Orleans.

Differs from the other species of the genus by having no excavation in the whorl behind the lip.

---

7. Polygyra polygyrella, Bland and Cooper.
Plate 11, figure 26.

Discoidal, shining, translucent; spire slightly elevated; whorls 7—8, ribbed above, smooth below; aperture armed with two rows of three teeth each, visible through the whorl, margins joined by a V-shaped tooth; base widely umbilicate, exhibiting about 3 whorls. Yellowish horn colored.

Diam. 11·5, alt. 5 mill.

Cœur d'Alêne Mountains.

It is very doubtful whether this species is properly placed in *Polygyra*; it differs in the teeth arranged in rows within the aperture.

Descriptions of additional species of Helices, and notes on some of those already described.

Aglaja sequoicola, Cooper.
Plate 11, figure 27.

" Shell rounded, umbilicate, spire depressed, last whorl sometimes subangulate, whorls 6 to $6\frac{1}{2}$, peristome oblique, little deflected above; labium thin, reflexed, thickest below; acute. Color dark brown or olivaceous, with a broad black band between two yellow ones, half hidden on the spire, lips white; within a fine purple with two white bands. Epidermis shining, polished below, the lines of growth faintly visible, sometimes very lightly malleated, and with spiral ridges; above with crowded scars bearing very short bristles in the young shell which fall off in the adult.

Animal slate colored, body cylindrical, rugose, tentacles moderate; foot elongated, behind wedge-shaped.

Shell—large diameter 0·96 to 1·20; smaller diameter 0·76 to 0·96; height 0·42 to 0·54 hundredths of an inch.

Santa Cruz, Cal., among decayed trees in the dampest places.

This beautiful species is quite rare, only nine adult and twelve young specimens having been found after long searching. It will probably occur more commonly in some part of the redwood forests which I have been unable to explore. It approaches nearest to *H. Dupetithouarsi* and *H. fidelis*, being between them in form and size as well as color, but the pilosity at once distinguishes it. Its distinct bands and rounded whorls separate it from *H. infumata* and *Hillebrandii*, the latter when perfect having also much longer hairs. The animal is lighter colored than those of *H. arrosa, Nickliniana, redimita, ramentosa, tudiculata* (which are all very similar), but much darker than that of *Dupetithouarsi*, and I believe also of *fidelis* and *infumata*. The form of the shell is a link connecting these with *Mormonum*."—*Proceedings California Acad. Nat. Sci., April*, 1866.

Through the kindness of Dr. Wesley Newcomb, of Oakland, California, I am enabled to give figures from his types of several species, which I was not able to illustrate at the time they were described.

Aglaja Ayresiana, Newcomb, sp. 7. Plate 11, fig. 28.

This species comes from the Island of Sta. Cruz; the locality originally given by Newcomb is incorrect. It is thus a subtropical, and not a boreal species.

Aglaja Bridgesii, Newcomb, sp. 11. Plate 11, fig. 29.

Aglaja Rowellii, Newcomb, sp. 18. Pl. 11, fig. 30.

Aglaja Gabbii, Newcomb, sp. 17. Plate 11, fig. 31.

I re-figure this species from a specimen received from Dr. Newcomb, the original figure, pl. 6, fig. 19, being unsatisfactory and more like the following species, to which *Gabbii* is closely allied.

Aglaja facta, Newcomb.

Plate 11, figure 32.

" Shell with perforation nearly covered, depressed orbicular, solid, compact, smooth, whitish, zoned with a brownish-red band; whorls 5 to 5½, somewhat convex, the last descending; suture slightly impressed; aperture oval; lip thick, reflected, yellowish.

Diam. 10, height 5 mill.

Islands of Sta. Barbara and San Nicolas, off the coast of California."—*Proc. California Acad. Nat. Sci.*, iii., 1864.

Polymita levis, Pfeiffer.

Plate 5, figure 21.

Plate 6, fig. 6, erroneously referred to this species, is a variety of *Arionta Veitchii*, Newcomb, from Cerros Island (not Bay of Monterey, Cal., as stated in description). The type of *Veitchii* is figured in pl. 5, fig. 19.

Conulus chersinella, Dall.

Plate 11, figures 33, 34, 35.

" Shell small, somewhat elevated, smooth, except that the lines of growth are occasionally indented; umbilicus minutely perforate; aperture semi-lunar and slightly oblique; whorls rotund, 4½ to 5 in number; sutures impressed, not deep; lip not thickened or reflected. Color yellowish, translucent.

Diam. ·14 in., height ·09 in.

Big Trees, Calaveras Co., California.

This small species has relations with *H. chersina*, Say, and *H. indentata*, Say. It resembles the former in its small size and many whorls, but differs in color and depressed spire, though sometimes almost as acute. It is related to the latter in its color and indented lines of growth, but differs in its greater number of whorls and much smaller size, and in the proportional size of the last whorl."—*American Jour. of Conchology*, ii. p. 328, 1866.

Hyalina Hornii, Gabb.

Plate 11, figures 36, 37, 38.

"Shell small, openly umbilicate, depressed; covered with an opaque brown epidermis, which, under the glass, shows minute oblique striations, and a few small, scattered hairs; whorls $4\frac{1}{2}$, the first $3\frac{1}{2}$, forming a very low, nearly flat spire, the last descending much more rapidly; suture strongly marked, especially between the last and penultimate whorl; umbilicus occupying about a third of the inferior surface, indistinctly perspective; aperture oblique, subcircular; lip simple, inner margins approximating.

Height ·09, diam. ·16 inch.

Fort Grant, junction of Arivapa and San Pedro Rivers, Arizona."—*Am. Jour. of Conchology*, ii. p. 330, 1866.

Gastrodonta significans, Bland.

Plate 11, figures 39, 40, 41.

"Shell umbilicate, depressed, discoidal, thin, with fine irregular striæ, which are almost obsolete at the base, shining, pale horn-colored, spire little elevated; suture slightly impressed; whorls 6, subplanulate, the last roundly inflated, rather flat at the base, excavated around the umbilicus, which is pervious, and equal almost to one-fifth the diameter of the shell; aperture oblique, depressed, lunate; peristome simple, acute.

Diam. 4·5, alt. 2 mill.

Fort Gibson, Indian Territory.

It is especially allied to *G. multidentata*, Binney, from which it differs in being of larger size, with wider umbilicus, and in the absence in the last whorl of the series of numerous small teeth which characterize Binney's species.

In a young specimen of *G. significans*, having four whorls only, there are, however, three small teeth, one by itself, and at some distance from it two others, situated as the teeth are in *G. multidentata*. Whether these teeth are or not constant in the ante-penultimate whorl of *G. significans*, I am unable to determine."—*Am. Jour. Conch.*, ii. p. 372, 1866.

Spurious Species.

HELIX HARPA, Say, belongs to the family Pupadæ.
HELIX IRRORATA, Say, = *H. lactea*, Müll., a Spanish species.
HELIX TRUMBULLI, Linsley, = *Skenea serpuloides*.

Family ORTHALICIDÆ.

Shell oval or elongated, with elevated spire, much longer than its width; aperture oval, entire below, the columella sometimes truncate at its termination; lip either sharp and simple, expanded or reflected, with or without teeth (none in North American genera), umbilicus generally covered.

Sub-families.

ACHATININÆ. Shell oblong; aperture oval, angulated above, rounded below, the lip sharp and not reflected, columella truncate below. Colors generally bright and variegated. Size large.

ORTHALICINÆ. Shell oblong, thin, imperforate; aperture oval, large, angulate above, rounded below, columella arcuate, thickened in the middle. Gaily painted in longitudinal reddish flames. Size moderate.

BULIMULINÆ. Shell oblong-turrited, moderately thick; aperture oval, small, outer lip generally expanded or reflected, inner lip reflected, axis perforate, rimate, or sometimes covered. Color white or brownish, sometimes variegated. Size small.

ACHATININÆ.

LIGUUS; Montfort.

Shell elongate-conical, spire elevated, apex acuminate, imperforate, solid, whorls 7—8, well rounded, the last about one-third of the total length; aperture semi-oval, margin thin, straight, columella obliquely subtruncate below. Gaily fasciate.

This is the only genus of the subfamily inhabiting the Western Continent, and the few species may be regarded as insular in origin, inhabiting principally Cuba, whence the two following have extended to the southern part of the adjacent coast of Florida.

1. Liguus fasciata, Müller.
Plate 12, figures 1, 2, 3. 5, 6.

Shell elongate-conical, striate by growth lines, solid, smooth, shining; spire elevated, apex acute, suture not deeply impressed; whorls 8, slightly convex, the last large, equalling from one-half to three-fifths of the total length of the shell; aperture semi-oval, generally pure white within, columella arcuate, and truncate at the base, with a rose-colored callus. Color white, variously ornamented with broad or narrow bands of yellow, green or purple, the apex, and sometimes the whole shell also, flamed with brown longitudinal zig-zags.

Length 55, diam. 25 mill.

Florida. (From Cuba.)

It is impossible to designate by description the extreme variation of coloring in this beautiful species; the variety with numerous green bands, and that with broad, yellow bands, (the latter the *Achatina solida* of Say,) are the most numerous in Florida.

---

2. Liguus picta, Reeve.
Plate 12, figure 4.

Shell ovate-conical, striate by growth lines, solid, smooth, shining; spire elevated conical, suture moderately impressed; whorls 7, the last large; aperture semi-oval, small, columella nearly perpendicular, truncate at base. Yellowish white, variegated externally by a double band of irregular brown spots upon the periphery, and above and below each sutural line, and also surrounding the columella, apex of spire and columella pink, whorls of spire with brown flames.

Length 44, diam. 24 mill.

Florida. (From Cuba.)

Differs from the foregoing principally in the pattern of coloring.

ORTHALICINÆ.

ORTHALICUS, Beck.

Shell ovate, imperforate, thin, striate, fasciate; whorls 6—8, the last inflated; aperture large, oval, lip thin and straight, columella sub-receding, obsoletely folded, rather thin.

This genus, together with the others comprising the subfamily, is of South American origin. In the northern part of that continent, the species are numerous, but one only inhabits the West Indies and the circumjacent North American coast. The *Orthalicus zebra*, distinguished by all American authors from the *undatus*, is scarcely even a variety of it, while the true *O. zebra* of Müller, a very different shell, inhabits the western parts of South America.

1. Orthalicus undatus, Ferussac.

Plate 13, figures 1, 2, 3.

Shell subconical, striated by growth lines, thick; spire elevated, suture moderate, slightly crenated: whorls 6, convex, the last about two-thirds of the total length of the shell; aperture large, ovate. White, with longitudinal undulated or zig-zag chocolate-colored flames, intersected by three narrow revolving lines of the same color; inner surface marked the same as the external.

Length 45, diam. 27 mill.

Southern Florida.

BULIMULINÆ.

All the genera of this subfamily are of South American origin, and only a few species of them extend into the subjacent parts of North America.

1. DRYMÆUS, Albers. Elongate-conical, perforate or rimate, thin, diaphanous, variegated; aperture large, oblong ovate, columella more or less twisted, peristome thin, expanded, columellar margin reflexed.

2. LIOSTRACUS, Albers. Oblong-conical, perforate, thin, smooth, fasciate; aperture obliquely semi-oval, lip thin, more or less expanded, the columellar margin dilated, reflexed.

3. MESEMBRINUS, Albers. Ovate-conical, rimate, rather thin, striate, white, generally variegated with brown; aperture oblong-oval, small, lip thin, and not reflected, columellar margin more or less dilated, reflected and appressed, columella slightly twisted.

4. THAUMASTUS, Albers. Shell oblong-conical, imperforate or rimate, nearly smooth, white, sometimes variegated with brown flames; aperture oblong-oval, lip obtuse, straight or slightly expanded, columellar margin reflexed, more or less appressed, columella distinctly twisted.

5. MORMUS, Albers. Shell oblong-conical, striate or subcostate, thin, whitish or variegated with brown, upper whorls flattened, the body whorl very convex, inflated; aperture subovate, the lips simple, sharp, columellar margin dilated and reflected.

6. SCUTALUS, Albers. Ovate-conical, perforate or umbilicate, roughly striate, whitish or brownish white, seldom variegated, last whorl ventricose, compressed at the base; aperture ovate-oblong, peristome more or less expanded, slightly thickened within.

7. PERONÆUS, Albers. Oblong-turrited, perforate, white, sometimes variegated with brownish, spire elevated, the last whorl one-third of the total length; aperture oblong-oval, lip expanded, not thickened, columellar margin dilated; columella receding or obsoletely arcuate.

DRYMÆUS, Albers.

1. Drymæus serperastrus, Say.

Plate 13, figure 4.

Shell ovate-fusiform, umbilicate, thin, translucent; spire acuminate, suture moderately impressed; whorls about 7, a little convex; aperture elongate-lunate, the lip expanded a little, and reflected upon its columellar margin; umbilicus moderate. Yellowish white, with about six interrupted bluish black bands on the body whorl, which sometimes coalesce; the internal coloring is the same, except near the lip margin, where the bands disappear.

Length 37, diam. 17·5 mill.

Texas and Mexico.

2. Drymæus Mexicanus, Lamarck.

Plate 13, figure 5.

Ovately acuminate, narrowly umbilicate, thin, pellucid, with thin incremental striæ. White with two brown zones on the last whorl, and maculations of the same color on the others.

Length 28 mill.

Cinaloa, North-western Mexico.

LIOSTRACUS, Albers.

1. Liostracus Ziegleri, Pfeiffer.

Plate 13, figure 6.

Shell ovate-conical, narrowly perforate, slightly striate, decussated by nearly obsolete spiral lines; spire conical, acute; whorls 6, slightly convex, the last subangulate in the middle; aperture oval, the lip simple, slightly reflexed on the columellar margin, columella scarcely receding. White, sometimes with chestnut bands, and interruptedly maculate upon the spire.

Length 21, diam. 10 mill.

Cinaloa, North-western Mexico.

---

2. Liostracus Floridanus, Pfeiffer.

Plate 13, figure 7.

Ovately turrited, perforate, rather smooth, hyaline with white opaque lines and maculations; spire elongate, acute; whorls $6\frac{1}{2}$, slightly convex, the last scarcely one-half of the total length of the shell, subangulate below the middle, and attenuated at the base; aperture oval, oblique, columella receding, a little twisted, columellar margin of the lip expanded and reflected. Interruptedly fasciate with brown.

Length 16, diam. 8 mill.

Florida.

3. Liostracus Dormani, W. G. Binney.
Plate 13, figure 8.

Ovately turrited, perforate, smooth; spire elongate, acute, suture impressed; whorls 6, slightly convex, with minute revolving lines, last whorl convex, with a very obtuse carina on the periphery; aperture semi-oval, columella perpendicular. Shining white, with several revolving rows of perpendicular, reddish-brown patches.

Length 29, diam. 12 mill.

St. Augustine, Florida.

MESEMBRINUS, Albers.

1. Mesembrinus multilineatus, Say.
Plate 13, figures 11, 12.

Ovate-conic, smooth; spire elevated, suture distinct, but not deep; whorls 7, slightly convex, the last three-fifths of the total length; aperture small, oval, columella perpendicular, perforation partly covered. Yellowish-white, with chestnut longitudinal lines, a dark infra-sutural line, and a black apex, umbilical area and lip.

Length 17, diam. 8 mill.

East Florida.

The markings are not unlike those of *M. virgulatus*, Fer., of West Indies.

---

2. Mesembrinus Humboldti, Reeve.
Plate 13, figure 13.

Ovately turrited, thin, smooth, narrowly umbilicated; spire elongate, acute; whorls 7, somewhat convex, the last three-fifths of the total length; aperture oblique, oval, lip sharp, not reflected, columellar margin dilated and appressed. White or yellowish, with narrow, interrupted brown bands, sometimes not banded.

Length 31, diam. 15 mill.

Cinaloa, North-western Mexico.

This species is described from Peru, and the identity of the Mexican specimens may be regarded as questionable.

3. Mesembrinus inscendens, Binney.

Plate 14, figure 21.

Shell rimate, thin, narrowly turrited, suture well marked: whorls 7, with minute revolving lines, the apex ribbed; aperture narrowly ovate, oblique, lip simple, columellar margin reflected. Reddish-brown.

Length 36, diam. 10 mill.

Lower California.

THAUMASTUS, Albers.

1. Thaumastus pallidior, Sowerby.

Plate 13, figure 9.

Elongate-ovate, rimate, striate by growth lines; spire elevated, acuminate, whorls 6, convex, the last two-thirds the total length; aperture subovate, lip reflexed, its extremities approaching and connected by a slight callus. White, yellowish white within.

Length 37, diam. 23 mill.

San Juan, Gulf of California; Cape St. Lucas, Lower California.

2. Thaumastus Californicus, Reeve.

Plate 13, figure 14.

Ovately turrited, thin, scarcely umbilicated; spire elevated-conical; whorls 6, smooth; aperture oval, lip sharp, expanded, columellar margin reflexed. White, with transverse bluish-black zones.

Length 19, diam. 10 mill.

California (Reeve).

Probably from Lower California. Very closely allied to *serperastrus*, Say.

3. Thaumastus excelsus, Gould.

Plate 13, figure 10.

Elongate ovate, acuminate, somewhat solid, smooth; spire elevated, acute; whorls 7, the last two-thirds the total length; aperture small sub-ovate, lip reflexed, columellar margin much expanded, lip extremities joined by a slight callus; axis rimate. Fulvous with white strigations, lip white.

Length 44, diam. 19 mill.

Lower California.

---

4. Thaumastus patriarcha, W. G. Binney.

Plate 13, figure 15.

Shell ovate, perforate, solid, rugosely striate; whorls 6, convex, the last ventricose and two-thirds of the total length; aperture ovate, lip thickened within, and its extremities joined by a heavy white callus, columellar margin slightly reflected over the umbilicus. White.

Length 35, diam. 19 mill.

Texas and Mexico.

Larger and more rugose than the allied species.

---

5. Thaumastus alternatus, Say.

Plate 13, figure 16.
Plate 14, figures 10, 12.

Ovate-conic, rather thick, umbilicated; suture slightly impressed; whorls 6, lip simple, expanded, thickened within, columellar margin reflected. White, yellowish or grey, with brown oblique longitudinal irregular or jagged bands, sometimes confluent.

Length 30, diam. 17 mill.

Texas and Mexico.

Fig. 16 represents the typical form, and fig. 12 a not quite adult, highly colored form; fig. 10 is a small and more numerous variety, if it be not indeed specifically distinct, It is the shell described by Menke as *Bulimus lactarius*.

6. Thaumastus Schiedeanus, Pfeiffer.

Plate 14, figures 1, 2, 4, 5.

Ovate-conic, thick, irregularly longitudinally striate, narrowly umbilicate; whorls $6\frac{1}{2}$, slightly convex, the last large; aperture oblong-oval, lip simple, a little expanded, thickened within, columellar margin reflected, columella more or less plicate. White, brownish inside.

Length 31, diam. 17 mill.

Texas and Mexico.

This species appears to vary considerably in form, some specimens being longer and less inflated than the type, and being obtusely angulated on the periphery. A young shell of this character is copied from W. G. Binney (fig. 2), who proposes, should it prove to be distinct, to call it *Mooreanus*. I add an older specimen, of the same form (fig. 4), but without the coloration of Mr. Binney's shell: it is from Brownsville, Texas. Fig. 5 represents a not unusual variety with distinct columellar tooth, which Pfeiffer supposed distinct, and proposed to call *Bulimus Binneyanus*. The examination of many shells showing the transition from a smooth to a toothed columella, convinces me that they are all of one species.

7. Thaumastus Mariæ, Albers.

Plate 14, figure 3.

Oblong conical, solid, rather smooth, narrowly umbilicate; spire conical, acute; whorls $6\frac{1}{2}$, slightly convex, attenuate at the base; aperture acuminately oblong, oblique, lip sharp, columellar margin dilated and reflected, columella with a small tooth. White, with obsolete brown spots and dashes, brown within.

Length 33, diam. 14 mill.

Texas.

MORMUS, Albers.

1. Mormus sufflatus, Gould.

Plate 14, figure 6.

Ovate, slightly striate, thin, slightly perforate; spire short; whorls $5\frac{1}{2}$, the last elliptical, and equalling three-fourths of the total length; aperture lunate, lip simple, columella reflexed. White.

Length 33, diam. 17 mill.

Lower California.

2. Mormus pilula, W. G. Binney.
Plate 14, figure 7.

Shell globular, thin, inflated, umbilicated; spire short-conical, suture well impressed; aperture rounded, lip thin, columellar margin broadly reflected. White, with two brown revolving bands.

Long. 12, lat. 9. mill.

Lower California.

SCUTALUS, Albers.

1. Scutalus proteus, Broderip.
Plate 14, figure 8.

This is a Peruvian species, of which a number of young specimens have been collected in Lower California. It may not be more than an adventitious inhabitant of that peninsula.

---

2. Scutalus dealbatus, Say.
Plate 14, figure 9.

Shell ovate-conical, thin, ventricose; whorls 7, striate, with growth lines which are more apparent on the spire, the last whorl subglobose; aperture ovate, lip thin, the columellar margin reflected, umbilicus narrow. White, with interrupted oblique longitudinal grey or yellow streaks.

Length 18, diam. 11 mill.

North Carolina, Alabama, Texas, Missouri, Arkansas.

A very abundant species, which will probably be found to inhabit all the far southern States.

---

3. Scutalus Xantusi, W. G. Binney.
Plate 14, figure 11.

Shell ovate-oblong, longitudinally striate, with minute revolving lines; spire elevated, suture impressed; whorls $5\frac{1}{2}$, slightly convex; aperture ovate, lip simple, columella arched, parietal wall covered by a slight callus. White (bleached?).

Length 21, diam. 8 mill.

Lower California.

PERONÆUS, Albers.

1. Peronæus artemesia, W. G. Binney.

Plate 14, figure 22.

Shell subcylindrical, rimate; whorls 8, gradually increasing in size, flattened, suture well impressed; surface smooth, except first whorl and a half of spire, which are ribbed; aperture small, obliquely oval, its margins approaching and connected by a heavy callous deposit. White, almost transparent.

Length 23, diam. 6 mill.

Lower California.

Addenda.

Bulimus (Leptomerus) Marielinus, Poey.

Plate 14, figure 23.

Ovate-conical, thin, imperforate; whorls 5, slightly convex, transparent, with several sub-interrupted brown bands on the lower portion of the body whorl; aperture oval, columella perpendicular, lip thin, not reflected.

Long. 8 mill.

A Cuban species, recently found in South Florida.

Spurious Species.

ACHATINA (COLUMNA) CALIFORNICA, Pfeiffer, quoted from Monterey, California, is a South American species.

BULIMUS DECOLLATUS, Linn.,
BULIMUS ACICULA, Müller,
BULIMUS SUBULA, Pfeiffer,
BULIMUS LUBRICUS, Müller,
BULIMUS GRACILLIMUS, Pfeiffer,
BULIMUS GOSSEI, Pfeiffer,
BULIMUS KIENERI, Pfeiffer,
BULIMUS HARPA, Say,
BULIMUS MARGINATUS, Say,

} Are all to be referred to the family Pupadæ.

Family PUPADÆ.

Shell cylindrical, with generally obtuse apex, the whorls numerous and nearly equal; aperture small, rounded, with expanded or reflected lips, and generally armed with teeth or laminæ within. Minute in size in most of the North American species.

Animal.—Tentacles very small or wanting; foot short, obtuse or pointed behind.

Remarks.—These shells are the smallest of all the terrestrial mollusca inhabiting the United States (except *P. incana*, which is a large species, but belongs to a West Indian group). They are so minute, indeed, that it requires the strictest scrutiny of the damp ground, moss, or decayed wood inhabited by them, in order to detect their presence, a difficulty which is much increased by their color, which is dull and earthy.

Genera.

1. CIONELLA, Jeffreys. Oblong-acuminate, smooth, polished; aperture small, oval, with a short, arcuate, more or less truncated columella.

2. STENOGYRA, Shuttleworth. Cylindrically turrited, generally truncate at apex, epidermis corneous, shining; aperture small, oval, columella truncate.

3. MACROCERAMUS. Turrited or conical, apex attenuated, last whorl angulated around its base; aperture oval, peristome not continuous nor reflected, except over the columella. Generally white with stripes or spots of darker markings.

4. PUPA, Draparnaud. Cylindrical, minutely perforate, size very small, aperture small, lip expanded or reflected, generally toothed within. Animal with superior and inferior tentacles.

5. STROPHIA, Albers. Shell large, with obtuse apex, the whorls generally covered with transverse ribs. Lip thickened and reflected, its extremities connected by a thick callus; columella dentate.

West Indian.

6. VERTIGO, Müller. Minute, cylindrical apex acuminate,' obtuse; lip expanded, white. Animal without inferior tentacles.

7. ZOOGENITES, Morse. Minute, turrited, conical, acutely ribbed longitudinally. Ovoviviparous.

CIONELLA.

Subgenera.

ZUA, Leach. Ovate-oblong, imperforate, smooth, pellucid, golden, shining; aperture small; lip obtuse, its extremities joined by a callus; columella more or less truncate.

ACICULA, Leach. Elongate, imperforate, white, vitreous, spire turrited, apex slightly obtuse; aperture oblong, lip simple, acute; columella subarcuate, truncate at base.

ZUA, Leach.

1. Zua subcylindrica, Chemnitz.

Plate 14, figure 14.

Shell oblong-oval, thin, polished, transparent; whorls 6, slightly convex, apex obtuse, sutures well marked; aperture oval, longitudinal, lip thickened but not reflected; umbilicus impervious. Bright amber color.

Length 7·5, diam. 2·5 mill.

Inhabits from New England and the Middle States to Lake of the Woods and the far Western Territories.

ACICULA, Leach.

1. Acicula acicula, Müller.

Plate 14, figure 13.

Cylindrical, acicular, spire attenuated, apex obtuse, suture narrowly margined; whorls 6 or 7, flat; aperture narrow, lip acute, columella arcuate. Hyaline, polished.

Length 4, diam. 1 mill.

European; introduced with imported plants, and sparingly distributed in a few places in the vicinity of greenhouses. Becoming naturalized.

This is an European species, introduced originally with imported plants; it has been found frequently in greenhouses, and lately, by Mr. A. D. Brown, of Princeton, N. J., in his garden.

STENOGYRA.

Subgenera.

RUMINA, Risso. Cylindrically elongate, truncate at apex, whorls remaining 4 or 5, smooth, shining; aperture semi-oval, lip margined within, not expanded, its extremities joined by a callus, columella not truncate; rimate.

OPEAS, Albers. Elongated, thin, striate, shining; whorls 6—8, the last compressed, perforate; aperture ovate oblong, equalling one-third to one-fourth of the total length, lip simple, acute, columellar margin reflexed.

MELANIELLA, Pfeiffer. Elongated, imperforate, costate; aperture ovate, effused at base, peristome simple, subcontinuous.

1. **Rumina decollata**, Linnæus.

Plate 14, figure 15.

Shell cylindrically turrited, rather thick, smooth, semitransparent; apex truncated, leaving 4 or 5 nearly flat whorls; aperture oval, angular above; outer lip thickened but not reflected, columellar lip reflected; umbilicus imperforate. Bright amber color.

Length (truncated) 25, diam. 10 mill.

Charleston, S. C.

This is a common European species, which has been introduced into various parts of the old and new world. In Charleston it is very numerous, though not found elsewhere in the United States.

2. Opeas subula, Pfeiffer.

Plate 14, figure 17.

Shell elevated, transparent; whorls 8, well rounded, the apex obtuse; aperture small, oval; base minutely perforated. White or yellowish.

Length 12, diam. 2·5 mill.

Florida (from Cuba).

---

3. Melaniella gracillima, Pfeiffer.

Plate 14, figure 16.

Elongated cylindrical, thin, with 20 to 30 sharp longitudinal ribs on each whorl; whorls 8, flattened, the suture deeply impressed, apex obtuse; the last whorl is angular below the middle; aperture small, elongated, oval, lip and columella both nearly perpendicular. White.

Length 7·5, diam. 1·6 mill.

Florida (from Cuba).

MACROCERAMUS, Guilding.

This genus belongs peculiarly to the West Indian fauna, and the species mentioned below are only stragglers from it into the subjacent parts of the United States.

---

1. Macroceramus Pontificus, Gould.

Plate 14, figure 20.

Fusiformly cylindrical, apex acuminate; whorls 9 to 12, slightly rounded, closely obliquely costulate; suture impressed, crenulate; aperture small, obliquely rounded, lip slightly reflected; base with a raised or carinated revolving line. White with brown and bluish clouds or bands crossing the whorls obliquely, and a colored band upon the carinæ.

Length 13, diam. 5 mill.

Florida.

2. Macroceramus Gossei, Pfeiffer.

Plate 14, figures 18, 19.

Fusiformly cylindrical, obliquely costulate; whorls 6, convex, sutures crenulate; aperture obliquely rounded, the lip slightly expanded; base subangulate, rimate. White with curved oblique dark bands and corneous dots; sometimes the markings are obsolete.

Length 11, diam. 3·6 mill.

Texas (Coll. Menke).

PUPA, Draparnaud.

Subgenera.

PUPILLA, Leach. Cylindrical, apex obtuse; whorls 5—9, corneous, somewhat shining; aperture rounded, lip expanded, scarcely reflected, armed with teeth within or without teeth.

LEUCOCHILA, Albers. Cylindrically ovate, apex somewhat obtuse, smooth, pellucid, shining; aperture semioval, edentulous, or armed with teeth or plications, lip thickened, reflexed; rimate.

Genus PUPILLA, Leach.

1. Pupilla badia, Adams.

Plate 15, figure 2.

Shell subcylindrical, with an obtuse apex; whorls about 6, moderately well rounded, with well-marked suture; aperture small, rounded, with a small tubercle on the parietal wall, and occasionally a tooth on the base of the lip; umbilicus perforate. Dark chestnut color.

Length 3, diam. 1·5 mill.

New England States.

It has been confounded with *P. muscorum*, of Europe, by some American and foreign naturalists, but differs in being larger with a less thickened lip.

2. Pupilla Hoppii, Möller.
Plate 15, figure 3.

Obtusely fusiform, whorls 5, very much rounded, with deep suture; aperture small, quadrately oval, with a small tooth on the parietal wall, and a tubercular thickening on the columellar portion of the lip; surface coarsely striate.

Greenland.

I have never seen this species, and the above description is made up from the only published figure of it. Müller's description is very short and meagre.

---

3. Pupilla Blandi, Morse.
Plate 15, figure 4.

Ovately cylindrical, apex obtuse; whorls 6, well rounded, with deep suture; aperture rounded, small, the lip subreflected, with a tooth on the parietal wall, one on the columellar portion, and a third remote within the base.

Length 3, diam. 1·5 mill.

Drift on Missouri River, near Fort Berthold.

---

4. Pupilla variolosa, Gould.
Plate 15, figure 5.

Ovately conical, apex obtuse, whorls 5, well rounded, with a profound suture; surface thickly and irregularly pitted with small round indentations, aperture small, obliquely oval, with lip slightly reflected, with a revolving lamellar tooth, a tooth on the columella and another on the base. Yellowish green.

Length 2 mill.

East Florida.

---

5. Pupilla pentodon, Say.
Plate 15, figure 6.

Ovately conical, with subacute apex, whorls 5, well rounded, with deep suture; aperture small, obliquely rounded, lip expanded, but not reflexed, with a white callous inner margin armed with two teeth on the columella, of which the upper one is largest, and from two to seven teeth on the outer lip, while one to two teeth are situated on the parietal wall. Light horn color.

Length 2 mill.

Maine to Ohio and southwards to Georgia.

The number of teeth developed on the lip increases with age from the minimum to the largest number mentioned above.

6. Pupilla decora, Gould.
Plate 15, figure 7.

Ovately cylindrical, apex obtuse; whorls 6, well rounded, with deep sutures; surface shining; aperture small, rounded, armed with 4 teeth, one on the parietal wall, one on the columella, the third on the base and the fourth on the outer lip; perforate; light amber color.

Length 2·5, diam. 1·3 mill.

Vicinity of Lake Superior.

7. Pupilla Rowelli, Newcomb.
Plate 15, figure 8.

Ovately fusiform, apex subacute; whorls 5, moderately rounded, with impressed suture; aperture small, oval, with a parietal, a basal, a columellar and an outer lip tooth; of these the basal is largest, and that on the parietal margin the next largest.

California.

8. Pupilla Californica, Rowell.
Plate 15, figure 9.

Ovately cylindrical, apex obtuse, whorls 5, slightly convex, with rib-like striæ; aperture quadrately oval, with a tooth on the parietal margin, another on the columellar, a third on the basal, and a fourth on the outer margin.

California.

Distinguished from *P. Rowellii* by its more obtuse outline, raised striæ and smaller basal tooth. Its striæ distinguish it principally from *P. decora*, Gould.

Genus LEUCOCHILA, Albers.

1. Leucochila marginata, Say.
Plate 15, figure 11.

Ovately turrited, spire rather obtuse, whorls 6, well rounded, smooth; aperture rounded, with widely reflected lip; perforate. Dark brown.

Length 6, diam. 2·5 mill.

Eastern, Middle and Western States.

More cylindrical than the following, with wide-margined white lip.

---

2. Leucochila fallax, Say.
Plate 15, figure 10.

Ovately turrited, apex acuminate; whorls 6, well rounded, smooth; aperture rounded, large, lip expanded but not reflected, without teeth; umbilicus perforate. Dark brown.

Length 5, diam. 2·5 mill.

Eastern and Middle States.

This species, described by Mr. Say as distinct from his *P. marginata*, has since been confounded with it. I was luckily able to point out the difference between them in Am. Jour. Conchology, i. p. 285.

---

3. Leucochila Arizonensis, Gabb.
Plate 15, figure 12.

Pupoid or cylindrical, with obtuse apex, suture well impressed; whorls 5½, convex, smooth, translucent; aperture suboval, edentate, lip thickened, strongly reflected, slightly emarginate near the posterior termination; imperforate. Corneous, lip white.

Length 5, diam. 2 mill.

Fort Grant, Arizona. Pike's Peak in the Rocky Mountains.

4. Leucochila hordacea, Gabb.

Cylindrical, apex obtuse; whorls 6, convex, with well impressed suture, smooth, thin; aperture small, edentulous; lip narrowly reflected and white; base umbilicate, the umbilicus bounded by an angle.

Length 2·8, diam. 1·1 mill.

Fort Grant, Arizona.

---

5. Leucochila modica, Say.

Plate 15, figure 14.

Ovately conical, fragile, spire acute; whorls 5, convex; aperture small, oval, lip turned over but not flattened, without teeth; imperforate. Pale horn color.

Length 2·5, diam. 1·5 mill.

Florida.

---

6. Leucochila armifera, Say.

Plate 15, figure 15.

Subfusiform, smooth, apex obtuse; whorls 6 to 7, with moderately impressed sutures; aperture small, oval, with widely reflected lip, much thickened within, its extremities nearly joined, connected by a callus deposit on the parietal margin; teeth generally four in number, namely, a parietal tooth, which is lamellar, large, and has one or more projecting points or is sometimes bifid, a rounded tooth on the left side, remote from the margin, and two others on the outer lip near its base; base of shell keeled, umbilicus perforate. Light horn color.

There are occasionally, in addition to the above, a distant tooth in the base of the aperture, and a marginal one near the top of the outer lip.

Length 5, diam. 2·5 mill.

Vermont to Kansas and southward to Kentucky.

7. Leucochila contracta, Say.
Plate 15, figure 16.

Subconical; whorls 6, with moderately impressed sutures; aperture small, trigonal, the margin continuous, reflected, much thickened within, the throat nearly filled by the large teeth which include an irregular one on the parietal margin, projecting into the aperture and concave on its right side, a tubercle on the umbilical side and another in the base, and a raised tooth on the outer margin of the lip; base keeled, minutely perforate. Light horn color.

Length 2·5, diam. 1·3 mill.

Maine to Iowa and southward to Florida and Texas.

---

8. Leucochila rupicola, Say.
Plate 15, figure 17.

Narrow, cylindrically fusiform, apex obtuse; whorls 6, slightly convex; aperture small, oval, the lip thickened within and widely reflected: teeth five in number, namely, one on the parietal wall, large and emarginate (sometimes deeply), a conical one, sometimes divided, on the umbilical margin, the third at the base, the fourth on the outer lip, and the fifth deeply seated within the outer lip; frequently all the teeth but the first two are wanting; umbilicus minutely perforate. Brownish.

Length 2·5, diam. 1·2 mill.

Pennsylvania to Florida, Arkansas.

---

9. Leucochila corticaria, Say.
Plate 15, figure 18.

Subcylindrical, apex obtuse, whorls five, convex, with well-impressed sutures; aperture small, rounded, with white, reflected lip, on the parietal wall is a small tooth, or rarely two teeth, and the umbilical side of the lip is also partially dentate, although occasionally both teeth are wanting. Substance thin and translucent, whitish,

Length 2·5, diam. 1·3 mill.

Northern, Middle and Western States; Mississippi.

10. Leucochila pellucida, Pfeiffer.
Plate 15, figure above 24.

Cylindrical, thin, shining, pellucid, apex obtuse; whorl five, convex, aperture semi-oval, five-toothed, one on the parietal wall, one on the columella, two on the middle of the outer lip, and a minute one on its base; lip simple, expanded. Light corneous.

Length 2, diam. 1 mill.

Texas ?

This is a Cuban species, quoted from Texas by Römer. It may be doubted whether the specimens collected by this gentleman were really the same as the Cuban shell.

STROPHIA.

1. Strophia incana, Binney.
Plate 15, figure 19.

Cylindrical, thick, opaque, apex obtuse; whorls eight to twelve, flattened, suture not deep; more or less heavily striate, sometimes almost smooth; aperture small, rounded-oval, lip white, thickened and reflected with a callus deposit bearing a tooth on the parietal wall and on the columella; base imperforate, carinate, with the striæ well developed. White, with sometimes irregular black flames or zig-zag markings.

Length 25—31, diam. 10 mill.

Florida and Cuba.

This is the only representative in our country of a numerous group of large species, inhabiting several West India Islands, but attaining their greatest development in Cuba.

VERTIGO.

1. Vertigo Bollesiana, Morse.
Plate 15, figure 25.

Ovate-cylindrical, apex obtuse; whorls four, well rounded, with impressed suture; aperture sub-trigonal, with a curved parietal tooth, two columellar teeth, the lower one the smallest, and two transverse lamelliform teeth within the outer lip and at its middle and base; lip thickened and subreflected; axis perforate.

Length 1·7, diam. ·9 mill.

Maine, New Hampshire, Massachusetts, New York, Norfolk, Va.

Smaller than *V. Gouldi*, which it much resembles; it differs in the outer lip not being smooth in the middle.

2. Vertigo corpulenta, Morse.
Plate 15, figure 24.

Ovate, striate, polished, translucent, apex obtuse; whorls four, very convex, with deep suture; aperture rounded, truncated by the parietal wall, lip slightly thickened and reflected; teeth four, small and obtuse, one on parietal wall, one on the columellar margin, one on the middle of the labrum and one near its base.

Length 2·5, diam. 1·3 mill.

Little Valley, Washoe Co., Nevada, Eastern slope of Sierra Nevada, alt. 6500 ft.

3. Vertigo Gouldi, Binney.
Plate 15, figure 20.

Ovate-cylindrical, apex obtuse; whorls four, convex, with well marked sutures; aperture semi-oval, truncated above by the parietal wall, the outer lip sub-reflexed, and incurved in the middle; there are two sharp teeth on the umbilical lip, two more within the outer lip and one on the parietal wall; umbilicus perforate. Chestnut color.

Length 2, diam. 1 mill.

New England and Middle States.

4. Vertigo milium, Gould.
Plate 15, figure 21.

Ovate with obtuse apex; whorls five, well rounded, with deep sutures; aperture semi-oval, truncated above, outer lip incurved in the middle, lip white and reflected, parietal wall covered with callus; there are two sharp teeth on the parietal wall, a broad tooth on the umbilical margin, with occasionally one or two little tubercles near its base, one in the base of the aperture and two on the outer lip; umbilicus open, rather large. Chestnut color.

Length ·90, diam. ·65 mill.

Inhabits nearly the whole country east of Mississippi River.

---

5. Vertigo ovata, Say.
Plate 15, figure 22.

Ovate conical, ventricose, apex conical, whorls five, very convex, with deep suture; aperture half round, truncate above, lip thickened within and reflected, marked externally by a groove, outer lip incurved in the middle; teeth six to eight, a large sharp one and a small one on the parietal wall, two on the columellar margin, one of them at its base, and two on the labrum, one of which is also basal; umbilicus open. Dark amber colored, shining.

Length 1·8, diam. 1 mill.

Northern United States.

---

6. Vertigo simplex, Gould.
Plate 15, figure 23.

Cylindrical, apex obtuse; whorls five, well rounded, with deep suture; aperture rounded, peristome nearly continuous, scarcely reflected, without teeth; umbilicus nearly covered by the lip. Dark brown.

Length 1·6, diam. ·8 mill

Maine to Pennsylvania.

---

7. Vertigo ventricosa, Morse.
Plate 15, figure 26.

Ovate-conic, smooth and shining, apex obtuse; whorls four, very convex, with deep suture; aperture small, rounded triangular, the middle of the outer lip incurved, lip widely reflected and thickened within; there is one large tooth on the parietal wall, another occupies the columellar lip, with a smaller one near its base, and there are two large teeth on the outer lip, making five in all; umbilicus smooth. Light chestnut color.

Length 1·6, diam. 1 mill.

Maine, New Hampshire, Mohawk and Greenwich, New York.

ZOOGENITES, Morse.

1. Zoogenites harpa, Say.
Plate 15, figure 1.

Ovate-conic, apex acute, thin, translucent, covered with sharp, thin, blade-like longitudinal ribs, apex acute and smooth, whorls four, convex, with impressed suture; aperture large, obliquely semi-circular, without teeth, lip not reflected; umbilicus minutely perforated. Light horn color.

Length 5, diam. 3·5 mill.

Maine to Iowa.

This is a boreal species, and has not been met with south of the above localities, I believe. It much resembles, except in size, *Helix Idahoensis*, Newcomb.

Family CYLINDRELLIDÆ.

Shell cylindrical, multispiral, truncate, frequently costate; aperture small, subcircular, edentuous, peristome expanded, continuous.

CYLINDRELLA, Pfeiffer.

Subgenera.

GONGYLOSTOMA, Albers.—Cylindrically fusiform, apex attenuate, costulate or striate, whorls 9—20, the last more or less protracted, obsoletely angulate; aperture circular, lip expanded.

HOLOSPIRA, Albers.—Shell turrited or fusiform, apex conical, not truncate; whorls 11—14, the last not protracted, base carinate; aperture subquadrangular, peristome expanded, columella plicate.

Genus GONGYLOSTOMA, Albers.

1. Gongylostoma Poeyana, Pfeiffer.
Plate 15, figure 27.

Elongate, thin, longitudinally sharply striate; spire very long, acuminate and truncate at the apex; whorls eleven, slightly convex; the last carinate at base; aperture rounded with continuous, acute lip. Corneous.

Length 15, diam. 4 mill.

Florida.

2. Gongylostoma jejuna, Gould.
Plate 15, figure 28.

Small, fusiformly elongate, solid, attenuate and truncated at apex; whorls about nine remaining, convex, with well impressed suture, the last whorl carinate at base; aperture campanulate, the lip continuous and not in contact with preceding whorl. Light horn color, with narrow longitudinal white lines.

Length 10, diam. 2·5 mill.

Florida.

3. Gongylostoma Coahuilensis, W. G. Binney.
Plate 15, figure 29.

Cylindrical, ventricose, thin, apex conically obtuse; whorls twelve, the upper ones smooth, the middle ones striate, the last two strongly ribbed, attenuated at base and not carinate; aperture subquadrate, with continuous peristome. White.

Length 29, diam. 7, mill.

Coahuila, Mexico.

Genus HOLOSPIRA, Albers.

1. Holospira Rœmeri, Pfeiffer.

Subcylindrical, apex obtusely conical, not truncate, substriate; whorls fourteen, flattened, the last carinate at the base, twisted downwards and free from the preceding whorl; the aperture with narrowly expanded lip, which is continuous, with a plica within. Pinkish white.

Var. β. smaller, more ventricose, whorls 12, the last less free.

Length 13—14, diam. 4·5 mill.

New Braunfels, Texas.

2. **Holospira Goldfussi**, Menke.
Plate 15, figure 31.

Elongate, swelled in the middle; apex conic, not truncated; thin, translucent, lightly striate; whorls twelve, flat, narrow, the last carinated at base; aperture obliquely sub-triangular, lid slightly expanded, continuous. Light horn color.

Length 11, diam. 4·4 mill.
Texas.

---

3. **Holospira Remondi**, Gabb.
Plate 15, figure 32.

Oblong-elliptical, subtruncated, whorls twelve, the first two and a half smooth, the rest densely obliquely striate, the last whorl convex at the base, aperture round, with reflected lip; umbilicus minute. White.

Length 11, diam. 3 mill.
Sonora, Mexico.

---

4. **Holospira Pfeifferi**, Menke.
Plate 15, figure 34.

Oblong-ovate, thin, pellucid, densely obliquely striate; spire oblong, conical, apex obtuse; whorls twelve, convex, subcompressed at the base; aperture rounded, peristome shortly reflected. White.

Length 17·5, diam. 5·6 mill.
Sonora, Mexico.

Larger, with fewer and larger striæ than *H. Rémondi*; the the whorls also are more rounded.

---

5. **Holospira irregularis**, Gabb.
Plate 15, figure 30.

Cylindrical, slender, whorls sixteen to eighteen, slightly convex, the last one subangulated below; suture slightly impressed; minutely umbilicate; aperture small, subquadrate, lip slightly expanded; surface with small longitudinal ribs. Color light brown.

Length 25, diam. 5 mill.
Lower California.

6. Holospira Newcombiana, Gabb.

Plate 15, figure 33.

Large, slender, tapering; whorls eleven and a half, flattened, the last subangular in the middle; aperture with widely expanded lip which is continuous; surface with fine irregular, undulating and occasionally broken ribs, radiately and obliquely disposed, the interstices crossed by microscopic revolving lines.

Length 1·9 inch, diam ·35 in.

Lower California.

Family LIMACIDÆ.

This Family includes the so-called naked snails or slugs, which possess a small shield-like shell concealed beneath the mantle of the animal.

The following Family, *Arionidæ*, possesses a few calcareous agglomerated granules only, instead of the shell; and the third family of Snails, the *Philomycenidæ*, have no vestige of a shell.

LIMAX.

Subgenera.

EULIMAX, *includes species Nos.* 1, 2, 3 and 4.

AMALIA, *includes species No.* 5.

---

1. Limax flavus, Linn.

Plate 16, figure 3.

Yellowish brown, covered with lighter colored longitudinally-disposed spots; surface granulate, the granules smaller and concentrically arranged upon the mantle; posterior termination acute and keeled. Base dirty white.

Length 3 to 4 inches when fully extended.

Seaports, (introduced from Europe) inhabiting cellars.

2. **Limax maximus**, Linn.
Plate 16, figure 2.

This very large French Snail has been found recently in cellars in Philadelphia, in such numbers as to warrant the belief that it has become a permanent addition to our mollusca. The figure is from a living specimen.

---

3. **Limax agrestis**, Müller.
Plate 17, figures 14, 15, 16.

Much smaller than the preceding; color varying from white through all shades of yellowish and grayish to brown or black, with or without small black dots, the mantle sometimes mottled with lighter color, white beneath. Upper surface somewhat rugose.

Length 1 to 2 inches.

Cities on the coast and their vicinity. (From Europe).

---

4. **Limax campestris**, Binney.
Plate 17, figures 11, 12, 13.

Cylindrical, elongated, thin; color black, lead or brownish, without spots; mantle finely concentrically striated, back prominently tuberculated.

Length one inch.

Northern and Western States. Indigenous.

---

5. **Limax Columbianus**, Gould.
Plate 16, figure 1.

Thick, carinated and obtusely pointed behind; foot wide, margining the body and furbelowed, with transverse oblique striæ, body longitudinally corrugated, mantle finely granulated; color dark greenish yellow, sometimes with large purplish black blotches.

Length five inches.

Neighborhood of Puget's Sound, Oregon.

Family ARIONIDÆ.

ARION.

Subgenera.

PROLEPIS. *Species No.* 1.
ARIOLIMAX, Mörch. *Species No.* 2.

1. Arion fuscus, Müller.
Plate 17, figures 9—10.

Light ash color, sometimes obscurely banded lengthwise; body cylindrical, narrow, its posterior extremity flat and rounded; upper surface longitudinally corrugated; head darker than the body, tentacles blackish.

Length one to two inches.

Maritime cities and their vicinity. (From Europe).

---

2. Arion foliolatus, Gould.
Plate 17, figure 1.

Reddish fawn color, coarsely obliquely reticulated with slate-colored lines; mantle concentrically marked with slate color, foot projecting around and forming a border to the body which is obliquely lineated; tentacles small and short.

Length three and a half inches.

Neighborhood of Puget's Sound, Oregon.

Family PHILOMYCENIDÆ.

TEBENNOPHORUS, Binney.

1. Tebennophorus Carolinensis, Bosc.
Plate 17, figure 6.

Whitish or yellowish-white, sometimes unmarked, but generally with clouds or spots of brown or black, forming three ill-defined longitudinal bands; tentacles blackish, surface longitudinally rugose.

Length three to four inches.

New England States to South Carolina, and westward to Missouri.

2. Tebennophorus dorsalis, Binney.

Plate 17, figures 7, 8.

Cylindrical, ashy blue with a black band on the middle of the back, tentacles black; surface minutely longitudinally rugose. the posterior termination acute.

Length 18 mill.

Found in woods. Vermont, near Boston, near Philadelphia, (Tryon).

Family VERONICELLIDÆ.

VERONICELLA.

1. Veronicella Floridana, Binney.

Plate 17, figures 2—5.

Elongate oval, extremities rounded; back arched, slightly wrinkled, dark ashy gray mottled with black, with a white line in the middle and an ill-defined black band at a little distance on each side of it; tentacles short, annulated, the lower ones indistinctly bifurcate; drab color below.

Length 56 mill.

Meta-lee-chee Key, Charlotte Harbor, West coast of Florida, under moist earth.

Family ONCHIDIIDÆ.

ONCHIDIUM.

1. Onchidium Carpenteri, W. G. Binney.

Plate 18, figure 39.

Lower California.

Family AURICULIDÆ.

Shell oval or oblong, generally of a dark brown or chocolate color, with a more or less long and narrow aperture, the lip of which is generally toothed within, not umbilicate; not operculate.

Animal amphibious; furnished with two tentacula, with eyes at their bases.

Remarks. Although air-breathers, these animals require a large amount of moisture, and are generally found on the sea-shore, in places covered by high tides. Both in the character of the animal and in that of the shell, they much resemble the fresh water pulmonifera, which they seem to connect with the strictly terrestrial species.

Sub-families.

AURICULINÆ. *Shell* ovate-oval, lip expanded or thickened.

Animal terrestrial, extending inland.

MELAMPINÆ. *Shell* oval or oblong, the outer lip *thin*, armed by a submarginal row of teeth within.

Animal inhabiting the sea-shore.

AURICULINÆ.

Genera.

ALEXIA, Leach. Ovate or oval, pointed, thin; aperture ovate, narrow, the lip broadly rounded and entire at the base; parietal wall with transverse laminæ; outer lip expanded, thickened within, and sometimes toothed.

CARYCHIUM, Müller. Pupiform, very thin, transparent, minute; aperture suboval, columella generally folded, parietal wall with one or two teeth; lip reflected.

ALEXIA.

1. **Alexia myosotis,** Draparnaud.

Plate 18, figures 1, 2.

Oval-elongate, thin, smooth and shining; spire produced, apex acute, suture distinct; whorls 7–8, slightly convex, the last about three-fourths of the total length of the shell; aperture long and narrow, lip thin, sometimes slightly dentate-ridged within, appressed at the base and slightly reflected over the minute perforated umbilicus. There is a sharp transverse tooth on the parietal wall, and a smaller one above it; columella folded. Dark horn color, the sutures narrowly banded with red.

Length 8, diam. 4 mill.

Sea coast. New England to New York.

CARYCHIUM.

1. **Carychium exiguum,** Say.

Plate 18, figure 3.

Oval-elongate, white, translucent, shining; spire long, apex obtuse; whorls 5 or 6, convex, very oblique; aperture oval, white, columella plaited in the middle, and slightly folded also near the base; lip thick, reflected; umbilicus perforated.

Length 1·6, diam. ·6 mill.

Northern, Middle and Western States, Arkansas, Texas.

MELAMPINÆ.

Genera.

1. MELAMPUS, Montfort. Conical-ovate with short obtuse spire; aperture long and narrow; inner lip folded several times, outer lip internally ridged, the ridge plicate. Foot of the animal bifid posteriorly.

2. TRALIA, Gray. Ovate with elevated spire; aperture long, narrow, dilated in front; inner lip 3-folded, outer lip numerously transversely ridged within. Foot of animal acute and entire behind.

3. LEUCONIA, Gray. Shell ovate-oblong, spire conical, aperture elongate, oval; inner lip with two anterior plaits, outer lip internally smooth, its margin acute. Foot divided inferiorly by a transverse groove.

4. PEDIPES, Adanson. Subglobose, imperforate, transversely striate, spire short; aperture narrow; parietal wall concave, with three plaits, of which the posterior is largest; outer lip with a sinus behind, two internal teeth, and an acute margin. Foot divided below by a transverse groove.

5. BLAUNERIA, Shuttleworth. Ovate-oblong, thin; aperture long and narrow, the inner wall plicate near the columella, columella subtruncate: lip simple, acute.

MELAMPUS, Montfort.

1. Melampus olivaceus, Carpenter.

Plate 18, figure 4.

Obconic; spire short, suture indistinct; whorls 7–9, obtusely angulated on the body below the suture; aperture long and narrow, lip covered with sharp laminæ within, parietal wall with from one to three small revolving laminæ; there is also a stout fold on the columella. Epidermis olivaceous, below which the color is white with patches or revolving lines of red.

Length 13, diam. 8 mill.

Lower California, Mazatlan.

---

2. Melampus bidentatus, Say.

Plate 18, figure 5.

Obovate, epidermis longitudinally wrinkled, with very minute revolving striæ; spire short, obtuse, suture distinct; whorls 6, the last obtusely angulated near the suture; aperture long and narrow, lip with or without laminæ within, when present sometimes terminating in an obtuse white callus running parallel with and slightly within the margin; parietal wall and columella each with a fold. Horn color or brownish, with generally four or five narrow reddish bands.

Length 10 mill., diam. 5·5 mill.

Whole Atlantic Coast of the United States.

---

3. Melampus flavus, Gmelin.
Plate 18, figure 6.

Obconic, smooth; spire short, convex, suture slightly impressed; whorls 9–10; aperture narrow, lip with about ten transverse ribs within, parietal wall and columella each with a prominent fold. Chestnut color, with three equidistant revolving light colored bands.

Length 13, diam. 8·6 mill.

Florida. (From West Indies.)

---

4. Melampus coffea, Linneus.
Plate 18, figures 7, 8.

Obconic, solid, smooth and shining, with microscopic revolving lines; spire short conic, apex black, pointed; whorls 9–10, the last angulated obtusely below the suture; aperture narrow, lip thickened within, with from 15–22 transverse laminæ, columella with a prominent fold, and two teeth on the parietal wall. Dark brown with three or four white revolving bands.

Height 19, diam. 10 mill., usually smaller.

Florida (from West Indies.)

TRALIA, Gray.
1. Tralia pusilla, Gmelin.
Plate 18, figure 9.

Ovate, solid, smooth, shining, with microscopic revolving lines; spire lengthened conic, with acute, black apex; whorls 6

—7, the last one slightly shouldered; aperture narrow above, much wider below, lip simple, acute, with a transverse lamina within, columella with a prominent fold, parietal wall with two teeth. Reddish-brown, with evanescent lighter bands, teeth white.

Length 11, diam. 5 mill.

Florida (from West Indies).

2. Tralia cingulata, Pfeiffer.

Plate 18, figure 10.

Fusiform, thick, polished, with microscopic revolving lines; spire elevated, convex-conic, apex acute, transparent; whorls 10, the last one not shouldered; aperture very narrow, lip acute, with 6 to 8 elongated laminæ not quite reaching to the edge, columella with a very prominent fold. Brown, with white revolving bands.

Length 11, diam. 5 mill.

Florida (from West Indies).

3. Tralia Floridana, Shuttleworth.

Plate 18, figure 11.

Obconic, inflated, thin, smooth; spire conic, apex acute; whorls 10, those of the spire radiately striate, body whorl obsoletely angled above; aperture narrow, lip acute, with transverse laminæ within, columella with a strong fold, parietal wall with two teeth. Gray, with chestnut bands.

Length 7·5, diam. 4·5 mill.

Florida Keys.

LEUCONIA.

1. Leuconia Sayi, Küster.

Plate 18, figure 12.

Ovate, thin, translucent, striate; spire elevated conical, apex acute; whorls 6, flattened, the last moderately convex; aperture small, oblong, lip sharp, columella biplicate, with occasionally a tooth on the parietal wall of old specimens. Corneous.

Length 5, diam. 3 mill.

This species is referred to the United States by Küster, but is unknown to American conchologists. Mr. Binney suggests that the specimen described may have been a variety of *Alexia myosotis*.

PEDIPES.

1. **Pedipes lirata**, W. G. Binney.

Plate 18, figure 38.

Shell globosely-conical, solid, with regular spiral lines; spire short, with obtuse apex; whorls 3, the upper ones small, the last equalling 5—6 of the total length; aperture semicircular; parietal wall with a strong transverse lamina, columella with two acute approximate teeth. White, or yellowish.

Length 3·3, diam. 2·5 mill.

Cape St. Lucas, Lower California.

BLAUNERIA.

1. **Blauneria pellucida**, Pfeiffer.

Plate 18, figure 13.

Shell reversed, ovate-elongated, pellucid, highly polished; whorls 7, the spire acuminate; aperture narrowly ovate, lip sharp, columella with a revolving fold.

Length 2·5, diam. ·8 mill.

Florida. Found also in a garden in Washington, D. C., where they are supposed to have been brought with plants from Charleston, S. C.

Family CYCLOPHORIDÆ.

Shell conical, elongate or depressed, varying greatly in the convexity of the whorls, as well as in solidity. Furnished with an operculum, the various methods of the accretion of which afford good generic characters.

Animal furnished with two tentacles, which are contractile but not retractile, with eyes at their external bases. Unisexual.

Remarks.—This family contains but one positively indigenous representative in the United States, although very numerous in species in the neighboring West Indian Islands. The presence of the operculum distinguishes the shell, as the contractile, non-retractile tentacles, with eyes at their external bases, do the animal from all preceding families.

Genus CHONDROPOMA.

1. Chondropoma dentatum, Say.

Plate 18, figures 14—16.

Conic, with 7 convex whorls, but generally truncate by the loss of three whorls; surface finely cancellated; suture deep and crenulated; aperture broadly ovate, a little angular above, lip continuous, slightly reflected; umbilicus small. Yellowish or brownish, with chestnut bands, sometimes interrupted so as to form longitudinal squares or stripes.

Length 12, diam. 4 mill.

Key West, Florida.

One of the figures represents this species suspended by a mucous thread, which it can spin at pleasure. It possesses this faculty in common with the naked slugs.

Family HELICINIDÆ.

Shell solid, depressed or lenticular, the whorls flattened, periphery frequently angulated; mouth half rounded, lip thick, generally reflected; umbilicus covered by a heavy deposit of callus. Operculum heavy, its growth annular.

Animal large, narrow; head extensile; tentacles narrow, with eyes on tubercles at their external bases.

Remarks.—This family is also, like the last, of West Indian extraction, where it flourishes greatly, the species being numerous, well marked, in most cases very beautiful, and varying greatly in size and ornamentation.

Genus HELICINA, Lamarck.

1. **Helicina orbiculata**, Say.

Plate 18, figures 17—19, 26.

Subglobose, not angulated; spire conical, apex acute; solid; whorls 5, with well impressed suture; aperture large, semilunar, lip reflected, and greatly thickened in old individuals. White (bleached), yellow, brown or gray, with frequently a pale or colored band on the periphery, and large or small, more or less numerous lines or spots.

Height 6, diam. 9 mill.

Tennessee, Mississippi, Georgia, Alabama, Florida, Texas.

A large, white variety, with greatly thickened lip, known as *H. tropica*, Jan., occurs plentifully in Texas. (Fig. 26.)

---

2. **Helicina occulta**, Say.

Plate 18, figures 20, 21.

Small, subglobose, depressed-conical, striated, solid; whorls 5, nearly plane, angular at periphery; aperture small, half round, lip thick. Yellowish or brownish.

Height 5, diam. 6·5 mill.

Found fossil and bleached in the post-tertiary of the Western States, Indiana, Ohio, Mississippi, etc. Mr. W. G. Binney mentions having received an apparently recent specimen from Shelboygan, Wis. Mr. Jacob Green, who described this same species subsequently to Mr. Say, gives us a locality " Hills, Western Pennsylvania." His specimens were undoubtedly recent. To this I have to add that at Lexington, Virginia, the species now exists numerously.

3. **Helicina Hanleyana,** Pfeiffer.

Plate 18, figures 22, 23.

Subglobose, rather solid, with somewhat distant concentric impressed lines; spire convexly conic, suture impressed; whorls 5, convex, the last well rounded; aperture semi-circular, lip a little expanded, thickened within. Reddish-brown, shining.

Height 5·6, diam. 7·5 mill.

Near New Orleans.

---

4. **Helicina chrysocheila,** Binney.

Plate 18, figure 24.

Elevated conical, or pyramidal, thin, shining, surface very finely punctured; whorls 5, convex, indistinctly angular at periphery, and the base somewhat flattened; aperture large, sub-oval, lip reflected. Flesh color or yellowish, interior of shell, lip and callus polished, deep golden color.

Height 8·3, diam. 10 mill.

Texas and Mexico.

---

5. **Helicina subglobulosa,** Poey.

Plate 18, figure 25.

Globosely conical, solid, lightly striate; whorls six, the last subangulate, base convex; aperture broadly semioval, lip wide, white, unicolor, or with two red bands, a broad one at the suture and a narrower one at the periphery.

Height 7, diam. 10 mill.

Florida, (from Cuba) at Key Biscayne; it is not known whether it was accidentally introduced, or is a permanent resident.

Family TRUNCATELLIDÆ.

Shell, cylindrical or pupoid, with small oval aperture and thin spiral operculum. Whorls transversely ribbed.

Animal with eyes at the rear of the base of the two contractile tentacles. Unisexual.

Remarks. Mr. W. G. Binney, in his Supplement to Terrestrial Mollusks remarks that extensive suites of the various Florida species of *Truncatella* show connecting links, which renders their separation exceedingly doubtful. I agree with him entirely, but as he has deferred to the opinions of Messrs. Pfeiffer and Poey, I do so likewise. The species of *Truncatella* all inhabit the vicinity of the sea.

1. Truncatella Caribæensis, Sowerby.
Plate 18, figures 27, 28.

Subcylindrical, rimate, solid, truncate; three or four whorls remaining, longitudinally ribbed, ribs slightly curved, sometimes becoming evanescent on the middle of the whorls, the last whorl sometimes smooth, carinate at the base; aperture obliquely ovate, peristome continuous, not reflected. Orange or red.

Length 7–8, diam. 3 mill.

Florida Keys.

2. Truncatella subcylindrica, Gray.
Plate 18, figures 29, 30, 31.

Subcylindrical, thin, pellucid, rimate, ribbed, ribs crowded, but sometimes evanescent, or visible near the suture only; whorls 4 remaining, somewhat convex but flattened in the middle, the last one somewhat carinated at base; aperture oblique, widely ovate, lip slightly thickened, its columella portion a little reflected. Light horn color.

Length 5, diam. 2 mill.

Florida Keys.

3. Truncatella bilabiata, Pfeiffer.
Plate 18, figures 32, 33.

Cylindrical, rimate, solid, opaque; ribs curved, elevated, obtuse; whorls remaining $4\frac{1}{2}$ to 5, convex, the last scarcely larger than the others; aperture vertical, broadly oval, scarcely angulated above; peristome double, the inner one continuous, the outer one heavy, white, terminating in a heavy basal carina. Brown.

Height 5–5, diam. 1·8 mill.

Florida Keys.

4. Truncatella pulchella, Pfeiffer.

Plate 18, figures 34, 35, 36.

Cylindrical, rimate, thin, pellucid, with thread-like, low, distant ribs, frequently evanescent; whorls $4-4\frac{1}{2}$, gradually increasing in size; the last generally smooth below the middle, base compressly carinate; aperture subvertical, ovate, lip simple, but somewhat expanded. Yellowish or reddish horn color.

Length 4·5–5, diam. 1·5–2 mill.

Florida.

---

5. Truncatella Californica, Pfeiffer.

Plate 18, figure 37.

Cylindrical, *imperforate*, thin, translucent, slightly striate; six to ten whorls, quite convex, the last one not carinate below; aperture vertical suboval, lip simple, continuous, slightly expanded. Amber color.

Length 4·6, diam. 1·6 mill.

San Diego, California.

LIST OF THE PRINCIPAL AMERICAN WORKS RELATING TO
TERRESTRIAL MOLLUSCA.

American Journal of Sciences and Arts. Published bi-monthly at New Haven, Conn., since the year 1817.
Contains papers by Conrad, H. C. Lea, Gould, etc.

Boston Journal of Natural History. Published by the Boston Natural History Society, since 1837.
Contains Binney's Monograph of Helices with illustrations. Vols. i to iv. This paper was afterwards expanded into the "Terrestrial Mollusks." Also papers by Anthony, W. G. Binney, Gould.

Annals of the Lyceum of Natural History. New York.
The most important paper is "Remarks on certain Species of North American Helicidæ, with Descriptions of New Species." By Thomas Bland. Vols. 6 and 7. Also published separately with the same title. Also papers by John H. Redfield, etc.

Thompson's History of Vermont. 1842.
Contains a paper on the Land and Fresh-water Shells, by Prof. C. B. Adams.

Proceedings of the Boston Society of Natural History. Vol. 1, 1843, and continued to present date; with papers in vols. 1 and 2, by Dr. Amos Binney, W. G. Binney, Gould, Leidy, Morse, Stimpson.

The Terrestrial Air-breathing Mollusks of the United States and Adjacent Territories of North America. By Dr. Amos Binney. Posthumous; edited by Dr. A. A. Gould. 2 vols., 8vo, of text. Boston, 1851. 1 vol. colored plates. Boston, 1859.

A Supplement to the Terrestrial Mollusks of the United States. By Wm. G. Binney. 8vo, with plates. Boston, 1859. (Published also in Boston Jour. Nat. Hist. vii.)

Proceedings of the Academy of Natural Sciences of Philadelphia. Published in 8vo, since 1842; containing important articles by W. G. Binney, Lea, Leidy, J. S. Phillips, Tryon.

Check List of the Terrestrial Gasteropoda. By W. G. Binney. Published by the Smithsonian Institution, Washington, D. C., 1860. Pamphlet.

Natural History of New York. Published by authority of the State. Part 5. Mollusca; by J. E. DeKay, M. D. 4to, pp. 227. 40 colored plates. Albany, 1843.

Invertebrata of Massachusetts (Report on). By Dr. A. A. Gould. 8vo, pp. 373. 1841. (Published by the State.)

Mollusca of Wilkes' Exploring Expedition. By Dr. A. A. Gould. 4to. Boston, 1852. Atlas of colored plates, in folio.

Lake Superior. By Prof. L. Agassiz. 8vo. Boston, 1850.
Contains descriptions of New Shells, by Dr. Gould.

Contributions of the Maclurian Lyceum. Philadelphia, 1827.
Contains Jacob Green's description of Helix Pennsylvanica.

Doughty's Cabinet of Natural History. 4to. Philadelphia.
Contains Jacob Green's description of Helicina rubella.

Transactions of the American Philosophical Society. New series. Philadelphia.
Contains the many valuable descriptions by Dr. Isaac Lea. Published also in 4to volumes, separately, with the title "*Obsertions of the Genus Unio,*" etc.

Journal of the Academy of Natural Sciences of Philadelphia. 1st series, in 8vo. From 1817 to 1842.
This Journal has the valuable papers of Thomas Say.

2d series, in 4to, to present date. With descriptions of several species by Dr. Lea.

Portland Journal of Natural History. 8vo, vol. 1. Portland, Maine.
Contains Prof. E. S. Morse's valuable paper on the Land and Fresh-water Mollusca of the State. Also published separately.

Nicholson's Encyclopedia. Three American editions—1816, 1818, 1819.
The article "Conchology," by Thomas Say, with descriptions of new species.

New Harmony Disseminator. New Harmony, Indiana (Weekly Newspaper). Descriptions by Thomas Say in 1829, 1830, 1831. Reprinted in pamphlet form with the title "Descriptions of some New Terrestrial and Fluviatile Shells of North America." New Harmony, 1840.

American Conchology, or Descriptions of the Shells of North America. By Thomas Say. Illustrated by colored figures. 8vo. New Harmony, Ind., 1830 to 1834.

Complete Writings of Thomas Say on the Conchology of the United States. Edited by W. G. Binney. 8vo. New York, 1858.

Descriptions of the Terrestrial Shells of North America, by Thos. Say. Edited by W. G. Binney. Pamphlet. Philadelphia, 1856.

Proceedings of the California Academy of Natural Sciences. Contains descriptions by Newcomb, Rowell, Dall, Cooper, etc.

American Journal of Conchology. Edited by Geo. W. Tryon, Jr. Philadelphia, 1865–6–7. With papers by Bland, Anthony, Gabb, Binney, Newcomb, Tryon, etc.

Land and Fresh-water Shells of the United States. By W. G. Binney. Part 3. (Limnophila and Operculata.) Published by the Smithsonian Institution, Washington, D. C. 8vo. 1866.

SHELLS FOR SALE.

The undersigned, contemplating expensive arrangements for the enlargement of his collection, will supply suites of

TERTIARY AND CRETACEOUS FOSSILS,

AND ALSO OF

LAND AND FRESH-WATER SHELLS

Of Alabama and adjacent States, at the rate of 30 cents a species, averaging six specimens to each species; representing, when practicable, the different ages of the land and fresh-water shells.

The more ponderous fossils will be represented by single specimens, or by pairs.

For further particulars communications are invited.

E. R. SHOWALTER,
UNIONTOWN, ALABAMA.

Works on Natural History,

FOR SALE OR EXCHANGE.

The following BOOKS, duplicates in my Library, are offered For Sale at the annexed prices, or will be Exchanged for Conchological Works. *GEORGE W. TRYON, Jr.,*
No. 625 Market Street, Philadelphia.

NATURAL HISTORY OF NEW YORK, 19 Vols., 4to., newly bound in half Morocco, with many hundred Colored Plates	$125 00
The following separate Volumes also for sale, in the same binding:	
MAMMALIA. By J. E. DeKay. 33 colored plates	5 00
ORNITHOLOGY. By J. E. DeKay. 380pp. 141 colored plates	20 00
HERPETOLOGY AND ICHTHYOLOGY. By J. E. DeKay. 2 Vols., 800pp. text, 101 colored plates	18 00
ENTOMOLOGY. By E. Emmons. 272pp. 47 colored plates	7 50
AGRICULTURE. By E. Emmons. 3 Vols., 770pp. 140 colored plates of Fruits, Vegetables,&c.	26 00
BOTANY. By J. Torrey. 2 Vols., 1000pp. 160 colored plates	25 00
GEOLOGY. 4 Vols., 1900pp. over 300 plates and maps	25 00
MINERALOGY. By L. C. Beck. 550pp.	6 00
CUVIER REGNE ANIMAL. Vol. 1, MAMMALIA, AVES. 8vo., 580pp., 1817	1 50
CRUSTACEANS, INSECTS, &c. 84pp., 1817	2 25
LINNÆUS SYST. NAT. 13th Edit. MAMMALIA, AVES. 8vo., 1032pp., 1788	1 75
AMPHIBIA, PISCES. 500pp., 1788	1 50
INSECTA. 2 Vols., 1500pp, 1788	3 00
MINERALS. 475pp., 1793	1 00
BROT CATALOGUE SYSTEMAT DES MELANIANS. 8vo., 72pp. Geneva, 1862	1 00
COUTHOUY, J. P. DESCRIPTIONS OF NEW SPECIES OF SHELLS OF MASSACHUSETTS BAY, MONOGRAPH OF OSTEODESMACEA, &c 118pp., 8vo., 3 plates	1 75
MIGHELS, J. W. CATALOGUE OF THE SHELLS OF MAINE. 8vo., 40pp., 1843	75
ENCYCLOPEDIE METHODIQUE. Zoophytes. By Lamouroux, St. Vincent and Deslongchamps. 4to., 1824. 2 Vols., 800pp. text, 100 plates	15 00
GRAY, LESSONS IN BOTANY. 8vo, cloth. 1859	50
BOTANY OF NORTHERN UNITED STATES, including MOSSES, &c. 8vo., half Morocco, 1858	1 50
GRAY, STRUCTURAL AND SYSTEMATIC BOTANY. 8vo., half Morocco. 1858	1 00

A MONOGRAPH

OF THE

TERRESTRIAL MOLLUSCA

OF THE

UNITED STATES.

With Illustrations of all the Species.

BY GEORGE W. TRYON, JR.

This work will be published in quarterly parts, each containing about 32 pp., 8vo., of text, and four lithographic plates crowded with figures. It is believed that about five parts will suffice to complete the work, which will then consist of about 155 pp. text, and 16 plates.

⁎⁎ Only 75 copies will be printed, so that an early application will be necessary to secure it. It has been thought advisable to issue the following different styles, 25 copies of each:—

    1st. Plain Edition. Printed on fine calendered paper, with uncolored plates.

    2d. Colored Edition. Same paper, the plates finely colored.

    3d. Fine Edition. On very heavy plate paper, and duplicate plates, plain on India paper, and colored.

The price has been fixed at the following low rate, which will not half cover the expense of publication:—

    Plain Edition, . . . $1.25 per Part.
    Colored " . . . 2.00 "
    Fine " . . . 3.00 "

    Cash payable on delivery.

Subscribers to the *American Journal of Conchology* will be supplied at 20 per cent. discount from the above prices.

Subscriptions received by

        GEORGE W. TRYON, JR.,
          625 Market Street,
            Philadelphia.

GLANDINA.

Synonymy and Reference to Plate 1.

Figs. 1, 2. G. TRUNCATA, Gmelin. Systema Naturæ, p. 3434. (1788.)   No. 1.
    Binney. Terr. Moll., ii., p. 301, t. 59, 60, (1851.)
    W. G. Binney, Terr. Moll., iv., p. 141, t. 80, f. 9, (1859.)
    Polyphemus glans, Say. Jour. Acad. Nat. Sciences, i, p. 282, (1818.)

Fig. 3. G. PARALLELA, Wm. G. Binney. Proc. Acad. Nat. Sciences, p. 189, (1857.)   No. 2
    G. parallela, W. G. Binney, Terrest. Moll., iv., p. 140, (1859.)
    G. truncata, var. Binney, l. c., p. 302, t. 62, f. 2, (1851.)

" 4. G. TEXASIANA, Pfeiffer. Proc. Zoological Soc., London, (1856.)   No. 3.
    Novitates Conchologicæ, viii., p. 82, t. 22, f. 11, 12, (1857.)
    W. G. Binney, Terrest. Moll., iv., p. 140, t. 77, f. 21, (1859.)
    G. truncata, var. Binney, l. c., p. 302, t. 61, f. 2, (1851.)

" 5. G. BULLATA, Gould. Proc. Bost. Soc. Nat. Hist., iii, p. 64, (Oct. 1848.)   No 4.
    Binney, Terrest. Moll., ii., p. 298, t. 62a, (1851.)

" 6. G. VANUXEMENII, Lea. Trans. Amer. Philos. Soc., v., p. 84, t. 19, f. 78, (1837.)   No. 5.
    Binney, l. c., p. 299, t. 62, f. 1, (1851.)

" 7. G. DECUSSATA, Deshayes. In Ferussac Hist., ii., p. 182, t. 123, No. 47.   No. 6.
    G. truncata, var. Binney, l. c., p. 302, t. 61, f. 1, (1851.)
    G. corneola, W. G. Binney. Proc. Acad. Nat. Sciences, p. 189, (1857.)
    W. G. Binney, Ter. Moll., Vol. iv., p. 139, (1859.)

Fig. 8. G. TURRIS, Pfeiffer. Symbolæ, iii., p. 91, (1846.) No. 7.
 Reeve, Conchologia Iconica, Achatina, t. 13, No. 45, (1849.)
 Carpenter, Catalogue of Reigen Collection, p. 175, (1857.)

" 9. G. ALBERSI, Pfeiffer. Proc. Zool. Soc. London, p. 295, (1854.) No. 8.
 Carpenter, Catalogue of Reigen Collection, p. 175, (1857.)

SUCCINEA.

SYNONYMY AND REFERENCE TO PLATE 2.

" 1. S. TOTTENIANA, Lea. Proc. Amer. Philos. Soc. ii., p. 32, (1841.) No. 1.
 Binney, Terrestrial Mollusks of the United States, t. 67*b*, fig. 2, (1857.)

" 2. S. INFLATA, Lea. l. c. p. 32, (1841.) No. 2.
 S. campestris, Say, var. Binney, l. c. ii., p. 66, (1851.)
 S. inflata, Lea. W. G. Binney, Terrest. Moll., iv., p. 34, t. 80, f. 11, (1859.)

" 3. S. UNICOLOR, Tryon. (Nov. species.) Am. Journ. Conch., vol. ii., No. 3, (1866.) No. 3.
 S. inflata, Lea, var. Pfeiffer, Monog. Hel viv., iii., p. 16, (1853.)

" 4. S. CAMPESTRIS, Say. Journ. Acad. Nat. Sci., i., p. 281, (1818.) No. 4.
 Binney, l. c. ii., p. 67*b*, fig. 1, (1851.)

" 5. S. STRETCHIANA, Bland. Annals N. Y. Lyceum of Nat. Hist., viii., (1865.) No. 5.

" 6. S. EFFUSA, Shuttleworth. Pfeiffer, Monog. Hel. Viv., iii., p. 17, (1853.) No. 6.
 W. G. Binney, l. c. iv., p. 41, t. 80, fig. 12, (1859.)

" 7. S. OBLIQUA, Say. St. Peter's Expedition, ii., p. 260, t. 15, f. 7. (1824.) No. 7.
 Binney, l. c. ii., p. 69, t. 67*b*, fig. 3, (1851.)
 S. ovalis, Say. Journ. Phila. Acad. Nat. Sci., i., p. 15, (1817.) ii., p. 163, (1821.)

S. lineata, DeKay. N. Y. Mollusca, p. 53, t. iv., f. 51, (1843.)
S. campestris, of American authors generally.

Fig. 8. S. GREERII, Tryon. (Nov. species.) Amer. Jour. Conchology, ii., No. 3, (1866.)   No. 8.

" 9. S. GROSVENORII, Lea. Proc. Acad. Nat. Sciences, p. 109, (1864.)   No. 9.

" 10. S. VERMETA, Say. New Harmony Disseminator, ii., No. 15, (1829.)   No. 10.
S. avara, Say. Binney, l. c., ii., p. 73, (1851.)
W. G. Binney, l. c., iv., p. 36, (1859.)

" 11. S. AVARA, Say. St. Peter's Exped., p. 260, t. 15, f. 5, (1824.)   No. 11.
Binney, l. c. ii., p. 74, t. 67c., fig. 4, (1851.)

" 12. *S. Wardiana*, Lea. Trans. Amer. Phil. Soc., ix., p. 3, (1844.)

" 13. S. GROENLANDICA, Beck. Pfeiffer, Monog. Hel. Viv., ii., p. 529, (1848.)   No. 12.
W. G. Binney, l. c. iv., p. 38, t. 80, fig. 4, (1859.)

" 14. S. GABBII, Tryon. (Nov. species.) Am. Jour. Conchology, ii., No. 3, (1866.)   No. 13.

" 15. S. VERRILI, Bland. Annals N. Y. Lyceum, viii., (1865.)   No. 14.

" 16. S. LINEATA, W. G. Binney. Proc. Acad. Nat. Sciences, ix., p. 19, (1857.)   No. 15.
Ter. Mollusks, iv., p. 39, t. 80, fig. 5, (1859.)

" 17. S. MOORESIANA, Lea. Proc. Acad. Nat. Sciences, p. 109, (1864.)   No. 16.

" 18. S. OREGONENSIS, Lea. Proc. Amer. Philos. Soc., ii., p. 32, (1841.)   No. 17.
Binney, l. c. ii., p. 77, t. 67c, fig. 2, (1851.)

" 19. S. RUSTICANA, Gould. Proc. Boston Soc. Nat. Hist., ii., p. 187 (1846.)   No. 18.
Mollusca of U. S. Expl. Exped., p. 28, fig. 29, (1852.)
W. G. Binney, Terr. Moll., iv., p. 6, t. 69, fig. 14, (1859.)

" 20. S. HAYDENI, W. G. Binney. Proc. Acad. Nat. Sciences, x., p. 114, (1858.)   No. 19.
Terr. Mollusks, iv., p. 40, t. 79, fig. 1, (1859.)

Fig. 21. S. Sillimani, Bland. Annals N. Y. Lyc. Nat. Hist., viii., (1865.) .... No. 20.

" 22. S. ovalis. Gould. Invertebrata of Massachusetts, p. 194, f. 125, (1841.)
Binney, l. c. ii, p. 78, t. 67a, fig. 3, (1851.) .... No. 21.

" 23. S. DeCampii, Tryon. (Nov. species.) Amer. Jour. Conchology, ii., No. 3, (1866.) .... No. 22.

" 24. S. Higginsi, Bland. (Nov. species.) Amer. Jour. Conchology, ii., No. 3, (1866.) .... No. 23.

" 25. S. retusa, Lea. Trans. Amer. Philos. Soc., v., p. 117, t. 19, fig. 86, (1837.) .... No. 24.
S. ovalis, Gld. (Part.) Binney, Terr. Moll., (1851.)
S. retusa, Lea. W. G. Binney, Terr. Moll., iv., p. 37, (1859.)

" 26. S. Nuttalliana, Lea. Trans. Am. Philos. Soc., ix., p. 4, (1844.)
Binney, l. c., ii., p. 81, t. 67a, fig. 2, (1851.) .... No. 25.

" 27. S. Wilsonii, Lea. Proc. Acad. Nat. Sciences, p. 109, (1864.) .... No. 26.

" 28. S. Forsheyi, Lea. Proc. Acad. Nat. Sciences, p. 109, (1864.) .... No. 27.

" 29. S. concordialis, Gould. Proc. Boston Soc. Nat. Hist., iii., p. 38, (1848.)
Binney, l. c., ii, p. 82, t. 67a, fig. 2, (1851.) .... No. 28.

" 30. S. luteola, Gould. Proc. Boston Soc. Nat. Hist., iii., p. 37, (1848.)
Binney, l. c., ii., p. 75, t. 67c, fig. 1, (1851.)
S. Texasiana, Pfeiffer, l. c., ii., p. 526, (1848.) .... No. 29.

" 31. S. Hawkinsii, Baird. Proc. Zool. Soc. London. No. 30.

" 32. S. Salliana, Pfeiffer. Proc. Zool. Soc. London, p. 133, (1849.)
W. G. Binney, l. c., iv., p. 42, t. 79, fig. 18, (1859.) .... No. 31.

" 33. S. aurea, Lea. Proc. Amer. Philos. Soc., ii., p. 32, (1841.)
Binney, l. c., ii., p. 76, t. 67c, fig. 3, (1851.) .... No. 32.

" 34. S. Haleana, Lea. Proc. Acad. Nat. Sciences, p. 109, (1864.) .... No. 33.

" 35. S. cingulata, Forbes. Proc. Zool. Soc., London, p. 56, t. 9, fig. 8, (1850.) .... No. 34.

HELICELLIDÆ.

SYNONYMY AND REFERENCE TO
PLATE 3.

VITRINA.

Fig. 1. V. LIMPIDA, Gould. Agassiz's Lake Superior, p. 243, (1850.)   No. 1.
Binney, Terr. Mollusks, ii., p. 58, t. 67a, fig. 1, (1851.)
V. Americana, Pfeiffer, Proc. Zool. Soc., London, p. 156, (1852.)
V. pellucida, (not of Müller,) DeKay, Moll., New York, p. 25, t. 3, figs. 4, 5, (1843.)

" 2. V. ANGELICÆ, Beck. Pfeiffer, Monog. Heliceorum Viv., ii., p. 510, (1848.)   No. 2.
W. G. Binney, l. c. iv., p. 33, t. 9, fig. 9, (1859.)
Helix pellucida, Fabricius, Fauna Grœnlandica, p. 389, (1780.)

" 3. V. PFEIFFERII, Newcomb. Proc. Cal. Acad. Nat. Sciences, p. 92, (1861.)   No. 3.

BINNEYA.

" 4. B. NOTABILIS, Cooper. Proc. Cal. Acad. Nat. Sciences, (1863.)   No. 1.

MACROCYCLIS.

" 5. M. NEWBERRYANA, W. G. Binney. Proc. Acad. Nat. Sciences, p. 115, (1858.)   No. 1.
W. G. Binney, l. c. iv., p. 20, t. 76, fig. 7, (1859.)

" 6. M. VANCOUVERENSIS, Lea. Trans. Am. Philos. Soc., vi., p. 87, t. 23, fig. 72, (1839.)   No. 2.
Binney, l. c. ii, p. 166, t. 20, (1851.)
H. vellicata, Forbes, Proc. Zool. Soc., London, p. 75, t. 9, fig. 1, (1850.)

" 7. M. SPORTELLA, Gould. Proc. Bost. Soc. Nat. Hist., ii., p. 167, (1846.)   No. 3.
Mollusca U. S. Expl. Exped., p. 37, fig. 42, (1852.)
Binney, l. c. ii., p. 211, t. 22a, fig. 1, (1851.)

Fig. 8. M. CONCAVA, Say. Jour. Acad. Nat. Sciences,
　　　ii., p. 159, (1821.)　　　　　　　　　　　　No. 4.
　　　Binney, l. c. ii., p. 163, t. 21. (1851.)
　　　　H. planorboides, Pfeiffer, Monog. Heliceorum,
　　　　　Viv. iii., p. 156, (1853.)
　　　　H. dissidens, Deshayes, Hist. Nat. des Moll., i.,
　　　　　p. 97, t. 84, figs. 1, 2.

" 9. M. VOYANA, Newcomb. Am. Jour. Conch., i.,
　　　p. 235, (1865.)　　　　　　　　　　　　　　No. 5.

" 10. M. ELLIOTTI, Redfield. Annals N. Y. Lyceum
　　　Nat. Hist., vi., p. 170, t. 9, figs. 8–10, (1856.)　No. 6.
　　　Binney, l. c. iii., p. 23, (1857.)
　　　W. G. B., l. c. iv., p. 116, t. 77, fig. 18, (1859.)

HYALINA.

" 11. H. INDENTATA, Say. Jour. Acad. Nat. Sciences,
　　　ii., p. 372, (1822.)　　　　　　　　　　　　No. 1.
　　　Binney, l. c. ii., p. 242, t. 29, fig. 2, (1851.)

" 12. H. FRIABILIS, W. G. Binney. Proc. Acad. Nat.
　　　Sciences, p. 187, (1857.)　　　　　　　　　　No. 2.
　　　Terrest. Mollusks, iv., p. 106, t. 80, fig. 2,
　　　(1859.)

" 13. H. LÆVIGATA, Rafinesque. Pfeiffer, Monog.
　　　Hel. Viv., i., p. 64, (1848.)　　　　　　　　No. 3.
　　　Binney, l. c. ii., p. 225, t. 32, (1851.)

" 14. H. LUCUBRATA, Say. New Harmony Dissemi-
　　　nator, ii., p. 229, (1829.)　　　　　　　　　　No. 4.
　　　Bland, Annals New York Lyceum, (1860.)

" 15. H. CADUCA, Pfeiffer, Zeitschr. für Mal., p. 146,
　　　(1846.)　　　　　　　　　　　　　　　　　　No. 5.
　　　Monog, Hel. Viv. i., p., 89, (1848.)
　　　Is not *H. lucubrata*, of Say.

" 16. H. FULIGINOSA, Griffith. Binney, l. c. ii., p. 222,
　　　t. 31, (1851.)　　　　　　　　　　　　　　　No. 6.

" 17. H. ARBOREA, Say. Nicholson's Encycl., iv., t.
　　　4, fig. 4, (1816.)　　　　　　　　　　　　　　No. 16.
　　　Binney, l. c. ii., p. 235, t. 29, fig. 3, (1851.)

" 18. H. SCULPTILIS, Bland. Annals N. Y. Lyc., vi.,
　　　p. 279, (1858.)　　　　　　　　　　　　　　No. 8.
　　　W. G. Binney, l. c. iv., p. 110, t. 77, fig. 15,
　　　(1859.)

Fig. 19. H. CELLARIA, Müller. Hist. Verm., No. 230, (1774.) — No. 9.
Binney, l. c. ii., p. 230, t. 29, fig. 4, (1851.)
H. glaphyra, Say? Nicholson's Encycl., iv., t. 1, fig. 3, (1816.)

" 20. H. CAPSELLA, Gould. Binney, l. c. ii., p. 239, t. 29a, fig. 1, (1851.) — No. 18.
H. rotula, Gould. (Pre-oc.) Proc. Bost. Soc. Nat. Hist., iii., p. 38, (1848.)

HELLICELLIDÆ.

SYNONYMY AND REFERENCE TO PLATE 4.

HYALINA.

" 21. H. KOPNODES, Wm. G. Binney. Proc. Acad. Nat. Sciences, p. 186, (1857.) — No. 7.
Terr. Mollusks, iv., p. 104, t. 80, fig. 14, (1859.)

" 22. H. INORNATA, Say. Jour. Acad. Nat. Sciences, ii., p. 371, (1822.) — No. 10.
Binney, l. c., ii., p. 227, t. 34, (1851.)

" 23. H. SUBPLANA, Binney, l. c., ii., p. 229, t. 33, (1851.) — No. 11.

" 24. H. NITIDA, Müller. Hist. Verm., ii., No. 234, (1774.) — No. 13.
H. lucida, Draparnaud. Hist. Nat. des Moll., p. 103, t. 8, figs. 11, 12, (1805.)
Binney, l. c., ii., p. 233, t. 22a, fig. 2, (1851.)
H. hydrophila, Ingalls' MSS.

" 25. H. ELECTRINA, Gould. Invertebrata of Mass., p. 183, fig. 111, (1841.) — No. 15.
Binney, l. c., ii., p. 236, t. 29, fig. 1, (1851.)

" 26. H. OTTONIS, Pfeiffer. Wiegmann's Archiv. fur Naturgesch., i., p. 251, (1840.) — No. 14.
Binney, l. c., ii., p. 238, t. 29a, fig. 3, (1851.)

" 27. H. BREWERI, Newcomb. Proc. Cal. Acad. Nat. Sciences, p. 118, (1864.) — No. 12.

" 28. H. VORTEX, Pfeiffer. Archiv. fur Naturgesch., ii., p. 351, (1839.) — No. 17.
H. selenina, Gould. Proc. Bost. Soc. Nat. Hist., iii., p. 38, (1848.)
Binney l. c. ii., p. 240, t. 29a, fig. 2, (1851.)

Fig. 29. H. ZONITES, Pfeiffer. Proc. Zool. Soc., London, p. 127, (1845.)

" 30. H. BILINEATA, Pfeiffer. Proc. Zool. Soc., London, p. 91, (1845.)

" 31. H. BINNEYANA, Morse. Portland Jour. Nat. Hist. i., (1864.)   No. 19.

" 32. H. FERREA, Morse. Portland Jour. Nat. Hist., i., (1864.)   No. 20.

MESOMPHIX.

" 33. M. INTERTEXTA, Binney, l. c., ii., p. 206, t. 36, (1851.)   No. 1.

" 34. M. LIGERA, Say. Jour. Acad. Nat. Sciences, ii., p. 157, (1821.)   No. 2.
Binney, l. c., ii., p. 204, t. 35, (1851.)
  H. Rafinesquea, Ferussac. Hist. Nat., t. 51a, fig. 5.
  H. Wardiana, Lea. Trans. Am. Philos. Soc., vi., p. 67, t. 23, fig. 82, (1839.)

" 35. M. DEMISSA, Binney, l. c., ii., p. 232, t. 42, fig. 1, (1851.)   No. 3.

" 36. M. CERINOIDEA, Anthony. Amer. Jour. Conchology, p. 351, t. 25, fig. 3, (1865.)   No. 4.

CONULUS.

" 37. C. CHERSINA, Say. Jour. Acad. Nat. Sci., ii., p. 156, (1821.)   No. 1.
Binney, l. c., ii., p. 243, t. 17, fig. 4, (1851.)

" 38. C. FABRICII, Beck, Pfeiffer. Zeit. fur Mal., p. 90, (1848.)   No. 2.
W. G. Binney, l. c., iv., p. 120, t. 77, fig. 17, (1859.)
See also figures 63, 64.

GASTRODONTA.

" 39. G. GULARIS, Say. Jour. Acad. Nat. Sciences, ii., p. 156, (1821.)   No. 1.
Binney, l. c., ii., p. 250, t. 37, figs. 3, 4, (1851.)
  H. bicostata. Pfeiffer. Monog. Hel. Viv., i., p. 182, (1848.)

Fig. 40. G. LASMODON, Phillips. Jour. Acad. Nat. Sci., viii., p. 182, (1842.) — No. 2.
  Binney, l. c., ii., p. 254, t. 37, fig. 2, (1851.)
  H. macilenta, Shuttleworth? Bern. Mit., p. 195, (1852.)

" 41. G. SUPPRESSA, Say. New Harmony Disseminator, ii., p. 229, (1829.) — No. 3.
  Binney, l. c., ii., p. 253, t. 37, fig. 1, (1851.)

" 42. G. INTERNA, Say. Jour. Acad. Nat. Sciences, ii., p. 155, (1821.) — No. 4.
  Binney, l. c., ii., p. 247, t. 30, fig. 4, (1851.)

" 43. G. MULTIDENTATA, Binney. l. c., ii., p. 258, t. 48, fig. 3, (1851.) — No. 5.

STROBILA.

" 44. S. LABYRINTHICA, Say. Jour. Acad. Nat. Sciences, i., p. 124, (1818) — No. 1.
  Binney, l. c., ii., p. 202, t. 17, fig. 3, (1851.)

" 45. S. HUBBARDI, Brown. Proc. Acad. Nat. Sciences, p. 333, (1861.) — No. 2.

ANGUISPIRA.

" 46. A. SOLITARIA, Say. l. c., ii., p. 157, (1821.) — No. 1.
  Binney, l. c., ii., p. 203, t. 24, (1851.)

" 47. A. ALTERNATA, Say. Nicholson's Encycl., 1st edit., t. 1, fig. 2, (1816.) — No. 4.
  Binney, l. c. ii., p. 212, t. 25, (1851.)
  H. scabra, Lamarck, Anim. Sans. Vert., vi., p. 288, (1822.)
  H. infecta, Pfeiffer, Mal. Blatt., p. 86, (1857.)
  H. strongylodes, Pfeiffer. Proc. Zool. Soc. London, p. 53, (1854.)

" 48. A. CUMBERLANDIANA, Lea. Trans. Amer. Philos. Soc., viii., p. 229, t. 6, fig. 61, (1843.) — No. 6.
  Binney, l. c., ii., p. 216, t. 31, (1851.)
  H. mordax, Shuttleworth, Bern. Mittheil, (1852.)

" 49. A. STRIGOSA, Gould. Proc. Bost. Soc. Nat. Hist. ii., p. 166, (1846.) — No. 5.
  Moll. U. S. Expl. Exped., p. 36, fig. 41, (1852.)
  Binney, l. c., ii., p. 210, t. 26, (1851.)

(*See also figs.* 52, 54.)

PATULA.

Fig. 50 P. PERSPECTIVA, Say. Jour. Acad. Nat. Sciences, i., p. 18, (1817.) No. 1.
Binney, l. c., ii., p. 256, t. 30, fig. 1, (1851.)
H. parvula, Deshayes, Encycl. Meth., ii., p. 217, (1830.)

" 51. P. STRIATELLA, Anthony. Bost. Jour. Nat. Hist., iii., p. 278, t. 3, fig. 2, (1840.) No. 2.
Binney, l. c., ii., p. 217, t. 30, fig. 2, (1851.)
(*See also fig. 53.*)

ANGUISPIRA.
(*See fig. 49.*)

" 52. A. COOPERI, Wm. G. Binney. Proc. Acad. Nat. Sciences, p. 118, (1858.) No. 3.
Terr. Moll., iv., p. 97, t. 77, fig. 11, (1859.)

PATULA.
(*See fig. 51.*)

" 53. P. DURANTI, Newcomb. Proc. California Acad. Nat. Sci., p. 118, (1864.) No. 3.

ANGUISPIRA.
(*See fig. 49.*)

" 54. A. IDAHOENSIS, Newcomb. Am. Jour. Conchology, ii., p. 1, t. 1, figs. 1, 2, 3, (1866.) No. 2.

PLANOGYRA.

" 55. P. ASTERISCUS, Morse. Proc. Bost. Soc. Nat. Hist., vi., p. 128, (1857.) No. 1.

PSEUDOHYALINA.

" 56. P. MILLIUM, Morse. Proc. Bost. Soc. Nat. Hist., vii., p. 28, (1859.) No. 6.
W. G. Binney, l. c., iv., p. 101, t. 79, figs. 4, 5, (1859.)

" 57. P. EXIGUA, Stimpson. Proc. Bost. Soc. Nat. Hist. iii., p. 175, (1850.) No. 5.
Binney, l. c., iii., p. 16, t. 77, fig. 19, (1857.)

" 59. P. CONSPECTA, Bland. Annals N. Y. Lyceum Nat. Hist., viii., (1865.) No. 4.

" 58. P. MAZATLANICA, Pfeiffer. Malak. Blatt., iii., p. 43, (1856.)
(*See also figs. 61, 62, 63.*)

HELICODISCUS.

Fig. 60. H. LINEATA, Say. Jour. Acad. Nat. Sciences, i., p. 18, (1817.)   No. 1.
Binney, l. c., ii., p. 261, t. 48, fig. 1, (1851.)

PSEUDOHYALINA.
(See fig. 59.)

" 61. P. INCRUSTATA, Poey. Memorias, i., p. 208, 212, t. 12, figs. 11–16, (1852.)   No. 3.
W. G. Binney, l. c., iv., p. 68,)1859.)
H. saxicola, (not of Pfeiffer,) Binney, l. c., ii., p. 174, t. 29a, fig. 4, (1851.)

" 62. P. MINUSCULA, Binney. l. c., ii., p. 221, t. 17a, fig. 2, (1851.)   No. 2.
H. minutalis, Morelet, Testacea Novissima, ii., p. 7, (1851.)
H. apex, Adams, Cont. to Conch., p. 36, (1849.)
H. Lavalleana, H. Mauriniana, D'Orb., Moll. Cuba, p. 161, t. 8, figs. 20–22, (1853.)

CONULUS.
(See figs. 37, 38.)

" 63. C. MINUTISSIMA, Lea. Trans. Amer. Philos. Soc., ix., p. 17, (1844.)   No. 4.
W. G. Binney, l. c., iv., p. 100, t. 77, figs. 6, 7, (1859.)
H. minuscula, Binney, ii., p. 221, (1851.)

" 64. G. GUNDLACHI, Pfeiffer. Wiegmann's Archiv. für Naturgesch., i., p. 250, (1840.)   No. 3.
H. egena, Gould, (not of Say,) in Binney, l. c., ii., p. 245, t. 22a, fig. 3, (1851.)

PSEUDOHYALINA.
(See Species 59.)

" 65. P. LIMATULA, Ward. Binney, l. c., ii., p. 219, t. 30, fig. 2, (1851.)   No. 1.

PATULA.
(See Species 53.)

P. WHITNEYI, Newcomb. Proc. Cal. Acad. Nat. Sciences, p. 118, (1864.)   No. 4.
P. CRONKHITEI, Newcomb, l. c., p. 180, (1865.)   No. 5.

HELICIDÆ.

Synonymy and Reference to Plate 5.

HYGROMIA.

Fig. 1. H. RUFESCENS, Pennant. British Zoology, fig. 34, (1776.) — No. 1.

" 2. H. HISPIDA, Linnæus. Systema Naturæ, p. 1244, Edit. Gmel., (1790.) — No. 2.

" 3. H. JEJUNA, Say. Jour. Acad. Nat. Sciences, ii., p. 158, (1821.) — No. 3.
   H. Mobiliana, Lea. Proc. Amer. Philosoph. Soc., ii., p. 82, (1841.)
   Binney, Terrest. Moll., ii., p. 172, t. 42, fig. 2, (1851.)

" 4. H. BERLANDIERIANA, Moricand. Memoires de Soc. de Histoire Nat. de Genève, vi., p. 537, t. 1, fig. 1, (1833.) — No. 4.
   Binney, Terrest. Moll., ii., p. 109, t. 49, fig. 1, (1851.)

" 5. H. GRISEOLA, Pfeiffer. Symbolæ ad Hist. Hel., i., p. 41, (1841.) — No. 5.
   W. G. Binney, Terr. Moll., iv., p. 50, t. 77, fig. 20, (1859.)
   H. albocincta et albozonata, Binney. l. c. i., p. 128, t. 2, (1851.)

AGLAJA.

" 6. A. INFUMATA, Gould. Proc. Boston Soc. Nat. Hist., v., p. 137, (1855.) — No. 1.
   Binney, l. c. iii., p. 13, (1857.)
   W. G. Binney, l. c., t. 79, fig. 2, (1859.)

" 7. A. HILLEBRANDI, Newcomb. Proc. California Acad. Nat. Sciences, p. 115, (1864.) — No. 2.

Fig. 8. A. FIDELIS, Gray. Proc. Zool. Soc., London, p. 67, (1834.)   No. 3.
W. G. Binney, l. c., p. 14, (1859.)
H. Nuttalliana, Lea. Trans. Amer. Philos. Soc., vi., p. 88, t. 23, fig. 74, (1839.)
Binney, l. c. ii., p. 159, t. 18, (1851.)

" 9. A. ANACHORETA, W. G. Binney. Proc. Acad. Nat. Sciences, Philadelphia, ix., p. 185, (1857.)   No. 4.
Terr. Moll., iv., p. 11, t. 76, fig. 5, (1859.)

" 10. A. ARROSA, Gould. W. G. Binney, l.c., p.15, t. 76, fig. 4, (1859.)   No. 5.
H. æruginosa, Gould. (Pre-oc.) Proc. Bost. Soc. Nat. Hist., v., p. 137, (1855.)
Binney, l. c. iii., p. 12, (1857.)

" 11. A. EXARATA, Pfeiffer. Proc. Zool. Soc., London, p. 108, (1857.)   No. 6.
W. G. Binney, l. c., p. 13, (1859.)

A. AYRESIANA, Newcomb. Proc. Cal. Acad. Nat. Sciences, p. 103, (1861.)   No. 7.

" 12. A. NICKLINIANA, Lea. Trans. Amer. Philos. Soc., vi., p. 100, t. 23, fig. 84, (1839.)   No. 8.
Binney, l. c., p. 119, t. 6a, not plate 6, (1851.)
W. G. Binney, l. c., p. 7, (1859.)

A. CARPENTERII, Newcomb. Proc. Cal. Acad. Nat. Sciences, p. 103, (1861.)   No. 9.

" 13. A. TUDICOLATA, Binney. Bost. Soc. Nat. Hist., iv., p. 360, t. 20, (1842.)   No. 10.
Binney, Terrest. Moll., ii., p.118, t.16, (1851.)
W. G. Binney, l. c., p. 7, (1859.)

A. BRIDGESII, Newcomb. Proc. Cal. Acad. Nat. Sciences, p. 91, (1861.)   No. 11.

" 14. A. MORMONUM, Pfeiffer. Proc. Zool. Soc., London, p. 109, (1857.)   No. 12.
W. G. Binney, l. c., p. 15, t. 79, fig. 21, (1859.)

" 15. A. RAMENTOSA, Gould. Proc. Bost. Soc. Nat. Hist., vi., p. 137, (1855.)   No. 13.
Binney, l. c. iii., p. 12, (1857.)

" 16. A. TRASKII, Newcomb. Proc. California Acad. Nat. Sciences, p. 91, (1861.)   No. 14.

Fig. 17. A. Dupetithouarsii, Deshayes. Revue Zool., p. 360, (1839.) No. 15.
  Binney, l. c. iii., p. 13, (1857.)
  W. G. Binney, l. c., p. 15, t. 76, fig. 9, (1859.)
  H. Oregonensis, Lea. Trans. Amer. Philos. Soc., vi., p. 100, (1839.)
    (*See also pl. 6, figs. 18, 19, 20.*)

ARIONTA.

" 18. A. Rémondi, Tryon. Proc. Acad. Nat. Sciences, Philadelphia, p. 281, t. 2, fig. 1, (1863.) No. 5.
" 19. A. Veitchii, Newcomb, MS. No. 1.
" 20. A. Californiensis, Lea. Trans. Amer. Philos. Soc., vi., p. 99, t. 23, fig. 79, (1839.) No. 2.
  Binney, l. c. ii., p. 121, t. 6, fig. 2, (1851.)
  H. vincta, Valenciennes, Voy. Venus Moll., t. 1, fig. 2.
    (*See also pl. 6, figs. 1 et seq.*)

POLYMITA.

" 21. P. levis, Pfeiffer. (See pl. 6, fig. 6.)

HELICIDÆ.

Synonymy and Reference to Plate 6.

ARIONTA.
(*Plate 5, figs. 18 et seq*)

" 1. A Kelleti, Forbes. Proc. Zool. Soc., London, p. 55, t. 9, fig. 2, a. 6, (1850.) No. 3.
  W. G. Binney, l.c., p. 17, t. 76, fig. 12, (1859.)
" 2. A. crebristriata, Newcomb. Proc. California Acad. Nat. Sciences, p. 116, (1864.) No. 4.
    (*See also fig. 17.*)

POLYMITA.

(See Plate 5, fig. 21.)

Fig. 3. P. TRYONII, Newcomb. Proc. Cal. Acad. Nat. Sciences, p. 116. (1864.)   No. 1.

" 4. P. INTERCISA, Wm. G. Binney. Proc. Acad. Nat. Sciences, Philadelphia, ix., p. 18, (1857.)   No. 2.
W. G. Binney, Terr. Moll., iv., p. 8, (1859.)
H. Nickliniana, var. Binney, l. c. ii., p. 120, t. 6, fig. 1 (middle,) (1851.)

" 5. P. AFEOLATA, Sowerby. Pfeiffer, Zeitschrift für Malak., p. 154, (1845.)   No. 3.
Binney, l. c. iii., p. 14, (1857.)
W. G. Binney, l. c., p. 19, t. 76, fig. 11, 3, (1859.)

" 6. P. LEVIS, Pfeiffer. Zeitschrift für Mal., p. 152, (1845.)   No. 6.
W. G. Binney, l. c., p. 18, t. 76, fig. 10, (1859.)

" 7. P. REDEMITA, Wm. G. Binney. Proc. Acad. Nat. Sciences, Philadelphia, ix., p. 183, (1857.)   No. 4.
Wm. G. Binney, Terr. Moll., iv., p. 9, (1859.)
H. Nickliniana, Lea, var. Binney, l. c. iii., t. 6, fig. 1, except middle figure, (1857.)

" 8. P. PANDORÆ, Forbes. Proc. Zool. Soc., London, p. 55, t. 9, fig. 3, *a, b*, (1850.)   No. 5..
Binney, l. c., p. 15, (1857.)
W. G. Binney, l. c., p. 18, t. 76, fig. 8, (1859.)
H. Damascenus, Gould. Proc. Bost. Soc. Nat. Hist., vi., p. 11, (1856.)

" 9–13. P. VARIANS, Menke. Pfeiffer, Monog. Heliceorum Viv., i., p. 238, (1848.)   No. 7.
W. G. Binney, l. c., p. 51, t. 78, fig. 22, (1859.)
H. polychroa et rhodocheila, Binney, l. c. ii., p. 128, t. 46, 47, (1851.)

TACHEA.

" 14–15. T. HORTENSIS, Müller. Hist. Vermium, ii., p. 57, (1774.)   No. 1.
Binney, l. c., p. 111, t. 8, (1851.)
W. G. Binney, l. c., p. 51, (1859.)
H. subglobosa, Binney. Journ. Bost. Soc. Nat. Hist., i., p. 485, t. 17, (1837.)

[xvi.]

POMATIA.

Fig. 16. P. ASPERSA, Müller. Hist. Vermium, ii., p. 59, (1774.)   No. 1.
    Binney, Terr. Moll., ii., p. 116, (1851.)
    W. G. Binney, l. c., p. 51, t. 77, fig. 4, (1859.)

ARIONTA.
(See figs. 1, 2.)

" 17. A. HUMBOLDTIANA, Valenciennes. Pfeiffer, Symbolæ, i., p. 37, (1841.)   No. 6.
    H. Buffoniana, Pfeiffer. Zeit. für Malakozoöl, p. 152, (1845.)
    Binney, l. c., p. 115, t. 43, (1851.)

AGLAJA.
(See Plate 5, figs. 6-17.)

" 18. A. RETICULATA, Pfeiffer. Malakozoöl, Blätter, p. 87, (1857.)   No. 13.
    W. G. Binney, l. c., p. 12, (1859.)

" 19. A. GABBII, Newcomb. Proc. California Acad. Nat. Sciences, p. 117, (1864.)   No. 17.

" 20. A. RUFOCINCTA, Newcomb. Proc. California Acad. Nat. Sciences, p. 117, (1864.)   No. 16.
    A. ROWELLII, Newcomb. Proc. California Acad. Nat. Sciences, p. 181, (1865.)   No. 18.

HELICIDÆ.

SYNONYMY AND REFERENCE TO
PLATE 7.

VALLONIA.

Fig. 1. V. PULCHELLA, Müller. Hist. Vermium. No. 322, 1774.

" 2. V. MINUTA, Say. Journal of Academy of Nat. Sciences, i., p. 123, Oct., 1817.
Dekay, Nat. Hist. N. York, Mollusca, p. 40, t. 3, f. 33, a. b. 1843.
Stimpson, Shells of New England, p. 54, 1851.
Morse, Portland Journal of Natural History, i. pt. 1, 1864.
H. pulchella, Müller. Binney, Terr. Mollusks, ii., p. 175, t. 17, f. 1, 1851.
W. G. Binney, Terr. Mollusks, iv., p. 69, 1859.   No. 1.

ULOSTOMA.

" 3. U. PROFUNDA, Say. Jour. Acad. Nat. Sciences, ii., p. 160, 1821.
Say, American Conchology, pl. 37, f. 3, Mar., 1832.
Binney, loc. cit., p. 177, t. 22, 1851.
W. G. Binney, l. c., p. 70, 1859.
H. Richardi, Ferussac. Hist. des Moll., t. 70.
Lamarck, Anim. Sans. Vert., vi., p. 72, Apr., 1822.   No. 1.

" 4. U. SAYI, Binney. Bost. Jour. Nat. Hist., iii., p. 579, t. 16, July, 1840.
Binney, Terr. Mollusks, ii., p. 180, t. 23, 1851.
W. G. Binney, l. c., p. 70, 1859.
H. diodonta,* Say. St. Peters' Exped., ii., p. 257, t. 15, f. 4, 1824.   No. 2.

* Preoccupied by Muhlfeldt, 1817.

MESODON.

Fig. 5, 6, 7. M. ALBOLABRIS, Say. Nicholson's Encyc., p. 181, t. 1, f. 1, 1816.
    Say, Am. Conch., t. 13, April, 1831.
    Binney, Terr. Moll., ii., p. 99, t. 2, 1851.
    Bland, Ann. N. Y. Lyc., vi., p. 359, Sept., 1858.
    W. G. Binney, l. c., p. 43, 1859.   No. 10.

" 8. M. EXOLETA, Binney. l. c., p. 121, t. 10, 1851.
    W. G. Binney, l. c., p. 54, 1859.
    H. zaleta, Binney. Bost. Jour. N. Hist., i., p. 492, t. 20, May, 1837.   No. 1.

" 9. M. DENTIFERA, Binney. Bost. Journ. Nat. Hist., i., p. 494, t. 21, May, 1837.
    Binney, Terr. Moll., ii., p. 134. t. 12, 1851.
    W. G. Binney, l. c., p. 55, 1859.   No. 2.

" 10. M. WHEATLEYI, Bland. Ann. N. Y. Lyc. Nat. Hist., vii., t. 4, f. 7, Dec., 1861.   No. 3.

" 11. M. CHRISTYI, Bland. Ann. N. Y. Lyc. Nat. Hist., vii., t. 4, f. 5–6, Dec., 1861.   No. 4.

HELICIDÆ.

SYNONYMY AND REFERENCE TO
PLATE 8.

MESODON.

Fig. 1. M. THYROIDES, Say. Jour. Acad. Nat. Sciences, i., p., 123, Oct., 1817, ii., p. 161, Jan., 1821.
    Say, Nicholson's Encyc., 1818.
    Say, American Conchology, p. 13, Apr., 1831.
    Binney, l. c., p. 129, t. 11, 1851.   No. 5.

Fig. 2. M. BUCCULENTA, Gould. Proc. Bost. Soc. Nat.
   Hist., iii., p. 40, June, 1848.
   Binney, l. c. iii., p. 9, t. 11, a. b, 1857.
   H. rufa, DeKay. Moll. N. York, p. 44, t. 3, f.
   30, a. b, 1843.
   Mitchener, Am. Jour. Conch., ii., p. 53, Jan.,
   1866. No. 6.

" 3. M. DEVIA, Gould, Proc. Bost. Soc. Nat. Hist.,
   ii., p. 165, Aug. 1846.
   Mollusca of Wilkes' Expl. Exped., p. 69, f.
   74, 1852.
   Binney, Terr. Moll., iii., p. 11, 1857.
   H. Baskervillei, Pfeiffer. Proc. Zool. Soc., 1849. No. 7.

" 4. M. ROEMERI, Pfeiffer. Roemer's Texas, p. 455.
   Pfeiffer, Zeitschrift für Malakol., p. 117, 1848.
   W. G. Binney, l. c., p. 55, t. 77, f. 3, 1859.
   H. dentifera, Pfeiffer. Monog. Hel. vivent., iii.,
   p. 269, 1853.
   Chemnitz, Conchyl. Cab., ii., p. 331, t. 131,
   f. 1–3. No. 8.

" 5. M. MAJOR, Binney. Bost. Jour. Nat. Hist., i.,
   p. 473, t. 12, May, 1837.
   Binney, Terr. Mollusks, ii., p. 96, t. 1, 1851.
   W. G. Binney, l. c., p. 43, 1857.
   H. albolabris. Ferussac. Hist. des Moll., t. 43,
   f. 4, t. 46a, f. 7.
   Bland, Annals N. Y. Lyc. Nat. Hist., vi., p.
   359, Sept., 1858. No. 9.

" 6, 7. M. TOWNSENDIANA, Lea. Trans. Am. Philos.
   Soc., vi., p. 99, t. 23, f. 80, 1839.
   Binney, Bost. Journ. Nat. Hist., iii., p. 371,
   t. 13, July, 1840.
   Binney, Terr. Moll., ii., p. 161, t. 19, 1851.
   Gould, Moll. Wilkes' Expl. Exped., p. 67, f.
   36, 1852.
   Newcomb, Am. Jour. Conchology, p. 343,
   1865. No. 15.

" 8. M. MULTILINEATA, Say. Jour. Acad. Nat.
   Sciences, ii., p. 150, Jan., 1821.
   Binney, Bost. Journ. Nat. Hist., i., p. 480, t.
   14, May, 1837.
   Binney, Terr. Moll., ii., p. 103, t. 3, 1851.
   W. G. Binney, l. c., p. 45, 1859. No. 14.

Fig. 9. M. PENNSYLVANICA, Green. Contributions to the Maclurian Lyceum, p. 8, Jan., 1827.
    Binney, Bost. Journal Nat. Hist., i., p. 483, t. 16, May, 1837.
    Binney, Terr. Moll., ii., p. 105, 7, 1851.
    H. Mitchelliana, Deshayes. in Ferussac, Hist., i., p. 137, t. 97, f. 4–7.   No. 11.

" 10. M. MITCHELLIANA, Lea. Am. Philos. Trans., vi., p. 87, t. 23, f. 71, 1836.
    Bland. Ann. Lyc. Nat. Hist. N. Y., vi., p. 339, Sept., 1858.
    W. G. Binney, Terr. Moll., iv., p. 47, 1859.
    H. clausa, (part) Binney. Terr. Moll., ii., p. 107, 1851.   No. 12.

" 11. M. DIVESTA, Gould. Binney, Terr. Moll., ii., p. 358, 1851.
    H. abjecta, (preoccupied.) Gould, Proc. Bost. Soc. Nat. Hist., iii., p. 40, Oct., 1848.
    Binney, Terr. Moll., ii., p. 122, 1851.   No. 13.

" 12, 13, 14. M. COLUMBIANA, Lea. Am. Philos. Trans., vi., p. 89, t. 23, f. 75, 1839.
    Binney, Terr. Moll., ii., p. 169, t. 5, 1851.
    W. G. Binney, l. c., p. 16, 1859.
    Newcomb, Am. Jour. Conch., i., p. 347, 1865.
    H. labiosa, Gould. Proc. Bost. Soc. Nat. Hist., ii., p. 165, Aug., 1846.
    Binney, l. c., p. 170, t. 13a., 1851.
    Gould, Moll. U. S. Expl. Exped., p. 67, f. 35, 1852.   No. 16.

" 15. M. DOWNIEANA, Bland. Annals N. Y. Lyc. Nat. Hist., vii., t. 4, f. 23–24, Dec., 1861. No. 17.

" 16. M. CLAUSA, Say. Jour. Acad. Nat. Sciences, ii., p. 154, Jan., 1821.
    Say, American Conchology, t. 37, f. 1, Mar., 1832.
    (Part) Binney, l. c., p. 107, t. 4, 1851.
    Bland, Ann. Lyc. Nat. Hist., vi., p. 336, Sept., 1858.
    W. G. Binney, l. c., p. 46, 1859.
    H. Pennsylvanica, (part) Pfeiffer. Monog. Heliceorum, i. p. 291, 1847.
    Chemnitz. Conch. Cab., ii., p. 51.
    Reeve, Conch. Icon. Helix., No. 676, Aug., 1852.   No. 18.

HELICIDÆ.

Synonymy and Reference to Plate 9.

XOLOTREMA.

Fig. 1. X. ELEVATA, Say. Jour. Acad. Nat. Sci., ii., p. 154, 1821.
    Say, Am. Conchology, t. 37, f. 2, Mar., 1832.
    Binney, Bost. Jour. Nat. Hist., i., p. 480, t. 19, May, 1837.
    Binney, Terr. Moll., ii. p. 126, t. 9, 1851.
    W. G. Binney, l. c., iv., p. 52, 1859.
    H. Knoxvillina, Ferussac. Hist. Nat. t. 49, f. 5, 6.
    H. Tennesseensis, Lea. Trans. Am. Philos. Soc., xix., p. 1, 1844.   No. 1.

" 2. X. CLARKII, Lea. Proc. Acad. Nat. Sci., x., p. 41, Mar., 1858.
    W. G. Binney, l. c. iv., p. 53, t. 77, f. 10, 1859.   No. 2.

" 3. X. OBSTRICTA, Say. Jour. Acad. Nat. Sci., ii., p. 154, 1821.
    Bland, Ann. N. Y. Lyc. Nat. Hist., vii., p. 438, 1862.
    H. palliata, var. *a.*, Say. Jour. Acad. Nat. Sci., ii., p. 152, 1821.
    Binney, Terr. Moll., ii., p. 136, t. 15, 1851.
    H. palliata, var. *Carolinensis*, W. G. Binney. l. c. iv., p. 57, 1859.
    H. Carolinensis, Lea, Trans. Am. Philos. Soc., iv., p. 108, t. 15, f. 33, 1831.
    H. helicoïdes, Lea. l. c. iv., p. 109, t. 15, f. 34, 1831.   No. 3.

" 4. X. PALLIATA, Say. Jour. Acad. Nat. Sci., ii., p. 152, 1821.
    Binney, Terr. Moll., ii., p. 136, *part*, t. 14, 1851.
    W. G. Binney, l. c. iv., p. 56, *part*, 1859.
    Bland, l. c. vii., p. 433, 1862.   No. 4.

See also fig. 7.

TRIODOPSIS.

Fig. 5. T. INTROFERENS, Bland. Ann. N. Y. Lyc.
Nat. Hist., vii., t. iv., f. 3–4, Apr., 1860. No. 3.
" 6, 13. T. TRIDENTATA, Say. Nicholson's Encyc.,
iv., t. 2, f. 1, 1816.
Binney, Terr. Moll., ii., p. 183, *part*, t. 27,
1851.
W. G. Binney, l. c. iv. p. 70, 1859.
Bland, l. c. vii., p. 423, 1862. No. 1.

XOLOTREMA.

" 7, 11. X. APPRESSA, Say. Jour. Acad. Nat. Sci.,
ii., p. 154, 1821.
Binney, Bost. Jour. Nat. Hist., iii., p. 356,
t. 8, July, 1840.
Binney, Terr. Moll., ii., p. 140, t. 13, 1851.
W. G. Binney, l. c. iv., p. 59, 1859.
Bland, l. c. vii., p. 432, 1862.
H. linguifera, Lamarck. Anim. S. vert. vi., p.
90, 1822. No. 5.

ISOGNOMOSTOMA.

" 8. I. RUGELI, Shuttleworth. Diagnosen neuer
Mollusken, No. 2, p. 18.
Bern, Mittheil., p. 198, 1852.
Binney, Terr. Moll., iii., p. 18, 1857.
W. G. Binney, l. c. iv., p. 60, t. 78, f. 15,
1859.
Bland, l. c. vii., p. 426, 1862. No. 2.

TRIODOPSIS.

" 9. T. HOPETONENSIS, Shuttleworth. Bern. Mittheil, p. 198, 1852.
Binney, Terr. Moll., iii., p. 17, 1857.
W. G. Binney, l. c. iv., p. 72, t. 77, f. 16,
1859. No. 4.

ISOGNOMOSTOMA.

" 10. I. INFLECTA, Say. Jour. Acad. Nat. Sci., ii.,
p. 153, 1821.
Binney, Bost. Jour., iii., p. 358, t. 9, f. 1,
July, 1840.

Binney. Terr. Moll., ii., p. 143, t. 45, f. 3, 1851.
W. G. Binney, l. c. iv., p. 59, 1859.
Bland, l. c. vii.. p. 425, 1862.
H. clausa, Ferussac. Hist. Nat., t. 51, f. 2.   No. 1.

TRIODOPSIS.

Fig. 12. T. FALLAX, Say. Jour. Acad. Nat. Sci., ii., p. 119, 1821.
DeKay, Moll. N. Y., p. 28, t. 3, f. 23, 1843.
W. G. Binney, l. c. iv., p. 71, 1859.
H. tridentata, (part) Binney. Terr. Moll., ii., p. 183, t. 28, 1851.   No. 2.

" 14. T. VULTUOSA, Gould. Proc. Bost. Soc. Nat. Hist., iii., p. 39, 1848.
Binney, Terr. Moll., ii., p. 189, t. 40a., f. 4, 1851.
W. G. Binney, l. c. iv., p. 75, 1859.
Bland, l. c. vii., p. 439, t. 4, f. 21, 1862.   No. 6.

" 15. T. MULLANI, Bland and Cooper. Ann. N. Y. Lyc. Nat. Hist., vii., p. 363, t. 4, f. 16–17, June, 1861.   No. 7.

" 16, 19. T. LORICATA, Gould. Proc. Bost. Soc. Nat. Hist., p. 165, Aug., 1846.
Binney, Terr. Moll., ii., p. 145, t. 29a., f. 2, 1851.
Gould, Moll. Wilkes' U. S. Expl. Exped., p. 68, fig 39, a. b. c., 1852.
W. G. Binney, l. c. ix., p. 14, 1859.
H. Lecontei, Lea. Trans. Am. Philos. Soc., x., p. 303, t. 30, f. 13, 1852.   No. 8.

" 17. T. YUCATANEA, Morelet. Testacea nov. Am. Centr., i., p. 9, 1849.   No. 5.

STENOTREMA.

" 18, 20. S. MONODON, Rackett. Linn. Trans., xiii., p. 42, t. 5, f. 2.
Binney, Bost. Jour. Nat. Hist., iii., p. 360, t. 10, f. 1, Jan., 1840.
Binney, Terr. Moll , ii., p. 147, t. 41, 1851.
W. G. Binney, l. c. iv., p. 60, 1859.
Bland, l. c. vii.. p. 431, Dec., 1861.

[xxiv.]

  H. fraterna, Say. St. Peter's Expedition, ii.,
   p. 257, t. 15, f. 3, 1824.
   Binney, Bost. Jour., iii., p. 363, t. 10, f. 2,
    Jan., 1840.
   H. convexa, Deshays. Lamarck, Anim. s. Vert.,
    viii., p. 112, 1838.   No. 1.

Fig. 21. S. STENOTREMA, Ferussac. Pfeiffer, Symb. ad.
   Hist. Hel., ii., p. 39, 1842.
  W. G. Binney, l. c: iv., p. 61, 1859.
  Bland, l. c. vii., p. 427, Dec., 1861.
  H. hirsuta, var. Binney. Terr. Moll., ii., p. 151,
   t. 42, f. 5, 1851.   No. 2.

" 22, 23. S. GERMANA, Gould. Binney's Terr. Mollusks, ii., p. 156, t. 40a., f. 3. 1851.
  Gould, Moll. Wilkes' Expl. Exped., p. 70, f.
   40, a. b. c., 1852.
  W. G. Binney, l. c. iv., p. 14, 1859.   No. 5.

" 24. S. HIRSUTA, Say. Jour. Acad. Nat. Sci., i., p.
   17, June, 1817.
  Say, Jour. Acad. Nat. Sci., ii. p. 161, Jan.,
   1821.
  Binney, Bost. Jour. Nat. Hist., iii., p. 365,
   t. 10, f. 3, July, 1840.
  Binney, Terr. Moll., ii., p. 150, t. 42, f. 3,
   1851.
  W. G. Binney, l. c. iv., p. 62, 1859.
  Bland, l. c. vii., p. 427, Dec., 1861.   No. 3.

" 25. S. LABROSA, Bland. Ann. N. Y. Lyc. Nat.
   Hist., vii., t. 4, f. 19.   No. 9.

" 26, 28, 29. S. SPINOSA, Lea. Trans. Am. Philos.
   Soc., iv., p. 104, t. 15, f. 35, 1834.
  Binney, Terr. Moll., ii., p. 154, t. 44, f. 1, 1851.
  W. G. Binney, l. c. iv., p. 65, 1859.
  Bland, l. c. vii.   No. 6.

" 27. S. EDGARIANA, Lea. Proc. Am. Philos. Soc.,
   ii., p. 31, Apr., 1841.
  Lea, Trans. Am. Philos. Soc., ix., p. 2. 1844.
  W. G. Binney, l. c. iv., p. 65, 1859.
  Bland, l. c. vii., pl. 4, f. 18.
  H. spinosa, var., Binney. Terr. Moll., ii., p. 155,
   t. 44, f. 2, 1851.   No. 7.

Fig. 31, 35. S. MAXILLATA, Gould Proc. Bost. Soc. N.
Hist., iii., p. 38, July, 1848.
Binney, Terr. Moll., ii., p. 157, t. 40a., f. 2,
1851.
W. G. Binney, l. c. iv., p. 65, 1859.   No. 4.

" 32, 33. S. BARBIGERA, Redfield.   Ann. N. Y. Lyc.
Nat. Hist., vi., p. 171, t. 9, f. 4, 5, 7, Dec.,
1856.
Binney, Terr. Moll., iii., p. 21, 1857.
W. G. Binney, l. c. iv., p. 63, t. 77, f. 2, 1859. No. 10.

" 34. S. EDWARDSII, Bland. Annals N. Y. Lyceum
Nat. Hist., vi., p. 277, t. 9, f. 14–16, Feb.,
1858.
W. G. Binney, l. c. iv., p. 63, t. 78, f. 7, 9,
1859.   No. 8.

HELICIDÆ.

SYNONYMY AND REFERENCE TO PLATE 10.

DÆDALOCHILA.

Figs. 1, 4. D. LEPORINA, Gould, Proc. Bost. Soc. Nat.
Hist., p. 39, 1848.
Binney, Terr. Moll., ii., p. 199, t. 40a., f. 1,
1851.
Bland, l. c. vi., p. 348, Feb., 1858.
W. G. Binney, l. c. iv., p. 92, 1859.
H. pustula, var., Pfeiffer.   Monog. Hel., iii.,
No. 1575, 1853.   No. 1.

" 2, 3. D. PUSTULOIDES, Bland. Ann. N. Y. Lyc.
Nat. Hist., vi., p. 350, Feb., 1858.
W. G. Binney, l. c. iv., p. 93, 1859.
H. pustula, Binney. Terr. Moll., ii., p. 201,
t. 39, f. 3, 1851.   No. 2.

Fig. 5, 36, 38. D. TEXASIANA, Moricand. Memoires
    Soc. Genev., vi., p. 538, t. 1, f. 1, 2, 1833.
  W. G. Binney, l. c. iv., p. 79, 1859.
  H. Tamaulipasensis, Lea. Proc. Acad. Nat.
    Sciences, p. 102, 1857.   No. 4.

" 6, 17. D. PUSTULA, Ferussac. Hist. Nat. des
  Moll., i., p. 78, t. 50, f. 1.
  Bland, l. c. vi. p. 346, Feb., 1858.
  W. G. Binney, l. c. iv., p. 94, t. 77, f. 12,
    1859.
  H. pustula, part, Binney. Terr. Moll., ii., p.
    201, 1851.   No. 3.

" 7, 9. D. THOLUS, W. G. Binney. Proc. Acad.
  Nat. Sciences, p. 186, 1857.
  W. G. Binney, Terr. Moll., iv., p. 81, t. 78,
    f. 21, 1859.   No. 8.

" 8. D. MOOREANA, W. G. Binney. Proc. Acad.
  Nat. Sci., p. 184, 1857.
  W. G. Binney, Terr. Moll., iv., p. 80, t. 78,
    f. 24, 1859.   No. 9.

" 10, 31. D. TRIDONTOIDES, Bland. l. c. vii., p. 424,
  t. iv., f. 11, 12, Dec., 1861.   No. 5.

" 11, 13. D. ACUTEDENTATA, W. G. Binney. Proc.
  Acad. Nat. Sci., p. 183, Oct., 1857.
  W. G. Binney, Terr. Moll., iv., p. 23, t. 76,
    f. 1, 1859.   No. 11.

" 12, 14. D. LOISA, W. G. Binney. Proc. Acad.
  Nat. Sci., p. 183, Oct., 1857.
  W. G. Binney, Terr. Moll., iv., p. 23, t. 76,
    f. 2, 1859.   No. 12.

" 15, 16, 18. D. ARIADNE, Pfeiffer. Zeitschr. für
  Mal., p. 120, 1848.
  W. G. Binney, Terr. Moll., iv., p. 76, t. 78,
    1, 3, 4, 1859.
  H. Couchiana, Lea, Proc. Acad. Nat. Sci., p.
    102, 1857.   No. 13.

" 19, 25. D. TROOSTIANA, Lea. Trans. Amer.
  Philos. Soc., vi., p. 107, t. 24, f. 119,
    1838.
  Bland, l. c., p. 288, t. 9, f. 21—23,
    Feb., 1858.
  W. G. Binney, l. c., p 88, t. 78, f. 11, 1859.

H. fatigiata, Binney, Boston Jour. Nat. Hist., p. 388, (Part. Excl. Syn.,) t. 19, f. 8, 1840.
Binney, Terr. Mollusks, ii., p. 193, (Part. Excl. Syn.,) 1852.
H. plicata, Binney, l. c. iii., t. 39, f. 2, 1857.   No. 17.

Fig. 20, 21. D. DORFEUILLIANA, Lea. Trans. Amer. Philos. Soc., vi., p. 107, t. 24, f. 118, 1838.
Bland, Annals N. Y. Lyc., vi., p. 294, t. 9, f. 24—26, Feb., 1858.
W. G. Binney, l. c. iv., p. 86, t. 78, f. 2, 14, 1859.
H. fatigiata, Binney, Bost. Jour. Nat. Hist., iii., p. 388, (Part. Excl. Syn. and Fig.,) 1840.
Binney, Terrest. Moll., ii., p. 193, (Part. Excl. Syn. et Fig.,) 1851.   No. 14.

" 22, 23, 26. D. FASTIGANS, L. W. Say. Bland, Ann. N. Y. Lyc., vii., 1859.
H. fatigiata, Say, Disseminator of Useful Knowledge, ii., p. 229, 1829.
Binney, l. c. ii., p. 193, (excl. syn., t. 39, f. 4, 1851.
Bland, l. c. vi., p. 283, Feb., 1858.
W. G. Binney, Terr. Mollusks, iv., p. 82, 1859.
H. Texasiana, var. *B.*, Chemnitz, edit. 2, p. 86, (excl. desc. syn. and fig.) 1846.
Pfeiffer, Monog. Helic., Viv. i., No. 1086, (excl. desc. et syn.,) 1848.
H. Dorfeuilliana, Deshayes, Ferussac, Hist. Nat. i., p. 73, t. 69, D. f. 3, (excl. syn).
H. Texasiana, Deshayes, Id., p. 74, (excl. desc. syn. and fig.)   No. 16.

" 24, 44. D. HINDSI, Pfeiffer. Proc. Zool. Soc., p. 132, 1845.
Binney, Terr. Moll., iii., p. 17, 1857.
W. G. Binney, l. c. iv., p. 92, t. 78, f. 5, 6, 8, 1859.   No. 7.

Fig. 27, 28, 29. D. Hazardi, Bland, Ann. N. Y. Lyc. Nat. Hist., vi., p. 291, Feb., 1858.
    W. G. Binney, l. c. iv., p. 84, t. 78, f. 13, 1859.
    H. plicata, Say, (preoccupied,) Jour. Acad. Nat. Sci., ii., p. 161, 1821.
    H. fatigiata, Binney, Bost. Jour. N. Hist., iii., p. 388, part, (excl. syn. and t. 619, f. 3,) 1840.
  Binney, Terr. Moll., ii., p. 193, (part, excl., pl. 39, f. 2,) 1851.
    H. Texasiana, Pfr., Monog. Hel., i., p. 418, (excl. desc. et syn.,) 1848.
    H. Dorfeuilliana, Deshayes, Ferussac's Hist. Nat. i., p. 73, (excl. desc. syn. and fig.)   No. 18.

POLYGYRA.

" 30, 37. P. Febigerii, Bland. American Jour. of Conchology, ii., p. 373, t. 21, f. 10, 1866.

DÆDALOCHILA.

" 32, 33, 34. D. Jacksoni, Bland. American Jour. of Conchology, ii., p. 371, t. 21, f. 8, 1866. No. 15

" 35, 39. D. ventrosula, Pfeiffer. Proc. Zool. Soc., p. 131, 1845.
    W. G. Binney, l. c. iv., p. 72, t. 77, f. 14, 1859.   No. 6.

" 40, 41, 43. D. Behrii, Gabb. American Jour. of Conchology, i., p. 208, t. 19, f. 5—9, 1865.   No. 10.

" 42. D. hippocrepis, Pfeiffer. Roëmer's Texas, p. 455, 1849.
    W. G. Binney, Terr. Moll., iv., p. 77, t. 78, f. 19, 1859.   No. 19.

HELICIDÆ.

SYNONYMY AND REFERENCE TO
PLATE 11.

DÆDALOCHILA.

Figs. 1, 2, 3. D. AURIFORMIS, Bland. Ann. N. Y.
Lyc., vii., p. 37, Dec., 1858.
    H. auriculata, Binney. Terr. Moll., ii.,
(part,) t. 40, f. 1, (right hand) 2, 1851.
    H. avara, Reeve. Conch. Icon., t. 121, No.
720, 1852.
    H. auriculata, Reeve. Conch. Icon., t. 119,
No. 700, 1852.   No. 20.

" 4, 5, 6. D. AVARA, Say. Nicholson's Encycl., 1st
Am. Edit., 1816.
Journal Acad. Nat. Sci, i., p. 277, 1818.
W. G. Binney, l. c. iv., p. 74, 1859.
Bland, Ann. N. Y. Lyc., vii., p. 30, Dec.,
1858.   No. 21.

" 7, 8, 9. D. ESPILOCA, Ravenel. Bland. Ann. N.
Y. Lyc., vii., t. 4, f. 12, Apr., 1860.   No. 22.

" 10, 11, 12. D. POSTELLIANA, Bland. Ann. N. Y.
Lyc., vii., p. 35, Dec., 1858.   No. 23.

" 13, 14. D. AURICULATA, Say. Nicholson's Encycl.,
Edit. 1, 1816.
Journal Acad. Nat. Sciences, i., p. 277, 1818.
Binney, Terr. Moll., ii., p. 186, (part,) t. 40,
Fig. 1, (left hand,) 1851.
Bland, Ann. N. Y. Lyc., vii., p. 26, Dec.,
1858.
W. G. Binney, l. c. iv., p. 73, 1859.   No. 24.

" 15, 16. D. UVULIFERA, Shuttleworth. Bern.
Mittheil., p. 199, Aug., 1852.
Chemnitz, 2d Edit., ii., p. 426, t. 148, f. 19,
20, 1853.

Binney, l. c. iii., p. 20, 1857.
W. G. Binney, Terr. Moll., iv., p. 75, 1859.
Bland, Ann. N. Y. Lyc., vii., p. 34, Dec., 1858.
H. florulifera, Reeve. Conch. Icon., No. 699, Aug., 1852.   No. 25.

POLYGYRA.

Figs. 17, 18. P. ANILIS, Gabb. American Journal of Conchology, i., p. 209, t. 19, f. 1—4, July, 1865.   No. 1.

" 19, 20, 21. P. CEREOLUS, Mühlfelt. Berlin Magazine, viii., p. 41, t. 2, f. 18, 1816.
Bland, Ann. N. Y. Lyc., vii., May, 1860.
(Part) W. G. Binney, l. c. iv., p. 90, t. 77, f. 23, 1859.
H. septemvolva, (Part) Binney, Terr. Moll., ii., p. 196, 1851.   No. 2.

" 22. P. SEPTEMVOLVA, Say. Nicholson's Encyc., 1st Edit., 1816.
(Part) Binney, l. c. ii., p. 196, t. 38, t. 39, f. 1, 1851.
Bland, Ann. N. Y. Lyc., vii., May, 1860.
H. planorbula, Lamarck. Ann. S. Vert., vi., p. 89.
H. cereolus, Mühlf., (Part) W. G. Binney, l. c. iv., p. 90, 1859.   No. 3.

" 23. 24. P. CARPENTERIANA, Bland. Ann. N. Y. Lyc., vii., May, 1860.
H. microdonta, Desh. W. G. Binney, l. c. iv., p. 91, t. 78, f. 23, 1859.
H. plana, Dunker. Phil. Icon., i., p. 51, t. 3, f. 11, 1845.   No. 4.

" 25. P. VOLVOXIS, Parreyss. Pfeiffer Symbolæ, iii., p. 80, 1846.
W. G. Binney, Terr. Moll., iv., p. 92, t. 78, f. 17, 1859.
H. septemvolva var., Bland. Ann. N. Y. Lyc., vii., 1860.   No. 5.

" 26. P. POLYGYRELLA, Bland and Cooper. Ann. N. Y. Lyc., vii., t. 4, f. 13—15, June, 1861.   No. 7.

AGLAJA.

(See also Plates 5 and 6.)

Fig. 27. A. SEQUOICOLA, Cooper. Proc. California Acad. Nat. Sciences, Apr., 1866.
" 28. A. AYRESIANA, Newcomb. (See reference to pl. 5.)
" 29. A. BRIDGESII, Newcomb. (See reference to pl. 5.)
" 30. A. ROWELLII, Newcomb. (See reference to pl. 6.)
" 31. A. GABBII, Newcomb. (See reference to pl. 6.)
" 32. A. FACTA, Newcomb. Proc. California Acad. Nat. Sciences, iii., 1864.

CONULUS.

" 33, 34, 35. C. CHERSINELLA, Dall. Am. Jour. Conch., ii., p. 328, t. 21, f. 4, Oct., 1866.

HYALINA.

" 36, 37, 38. H. HORNII, Gabb. Am. Jour. Conch., ii., p. 330, t. 21, f. 5, Oct., 1866.

GASTRODONTA.

" 39, 40, 41. G. SIGNIFICANS, Bland. Am. Jour. Conch., ii., p. 372, t. 21, f. 9, Oct., 1866.

ORTHALICIDÆ.

SYNONYMY AND REFERENCE TO
PLATE 12.

LIGUUS.

Figs. 1, 2, 3, 5, 6. L. FASCIATA, Müller. Hist. Verm., ii., p. 145, 1774.
Binney, l. c. ii., p. 266, t. 55, (not 56,) 57, 1851.

W. G. Binney, l. c. iv., p. 138, 1859.
Bulimus vexillum, Brug.
Achatina solida, Say. Jour. Acad. Nat. Sci., v., p. 122, 1825.
A. crenata, Swainson.
A. pallida, Swainson.
Bulla virginea, B. Linn. Syst. Nat., Edit. 12, p. 1186.
A. lineata, Valenciennes. Recueil d'Observ., p. 248, t. 55, f. 2, 1833.
A. murrea, Reeve. Conch. Icon., t. 7, f. 22, a, b.
A. anais, Lesson.
A. lutea, Weigmann.
Bulimus zebra, Orb. Moll. Cuba.   No. 1.

Fig. 4. L. PICTA, Reeve. Conch. Icon., t. 10, f. 34.
Bulimus fasciatus, (Part) Binney, l. c. ii., p. 266, t. 56, 1851.
W. G. Binney, l. c. iv., p. 138, 1859.   No. 2.

ORTHALICIDÆ.

Synonymy and Reference to Plate 13.

ORTHALICUS.

Figs. 1, 2, 3. O. UNDATUS, Ferussac. Hist. des Moll., p. 52, t. 115, f. 194, t. 114, f. 5, 6.
Shuttleworth, Notitiæ Malacol, p. 63, t. 3, f. 4, 5, 1856.
Bulimus zebra, (Not of Müller,) Binney, l. c. ii., p. 271, t. 54, 1851.
W. G. Binney, l. c. iv., p. 125, t. 77, f. 13, t. 78, f. 12, 1859.   No. 1.

DRYMÆUS.

" 4. D. SERPERASTRUS, Say. New Harmony Disseminator, Dec. 28, 1831.

    Binney, l. c. ii., p. 274, t. 50, f. 2, 1851.
    W. G. Binney, l. c. iv., p. 126, 1859.   No. 1.
Fig. 5. D. MEXICANUS, Lamarck. Anim. S. Vert., viii., p. 252, 1838.
    Reeve, Conch. Icon., t. 40, f. 244.
    Carpenter, Mazatlan Cat., p. 177, 1857.   No. 2.

LIOSTRACUS.

" 6. L. ZIEGLERI, Pfeiffer. Proc. Zool. Soc., p. 113, 1846.
    Reeve, Conch. Icon., t. 58, f. 389.
    Carpenter, Mazatlan Cat., p. 177, 1857.   No. 1.
" 7. L. FLORIDANUS, Pfeiffer, Zool. Proc., p. 330, 1856.
    W. G. Binney, l. c. iv., p. 134, t. 79, f. 3, 1859.   No. 2.
" 8. L. DORMANI, W. G. Binney. Proc. Acad. Nat. Sci., p. 188, 1857.
    Terr. Moll., iv., p. 132, t. 80, f. 10, 1859.   No. 3.

THAUMASTUS.

" 9. T. PALLIDIOR, Sowerby. Proc. Zool. Soc., p. 108, 1832.
    Reeve, Conch. Icon., t. 55, f. 365.
    Bul. vegetus, Gould. Bost. Jour., vi., p. 375, t. 14, f. 2.
    Gould, Otia Conchologica, p. 184, 1862.   No. 1.
" 10. T. EXCELSUS, Gould. Bost. Jour., vi., p. 376, t. 14, f. 3, Oct., 1853.
    W. G. Binney, l. c iv., p. 24, t. 79, f. 12, 1859.   No. 2.

MESEMBRINUS.

" 11, 12. M. MULTILINEATUS, Say. Proc. Acad. Nat. Sci., v., p. 120, 1825.
    W. G. Binney, l. c. iv., p. 132, 1859.
    Bul. virgulatus, (not of Fer.,) Binney, l. c. ii., p. 278, t. 58, 1851.   No. 1.
" 13. M. HUMBOLDTI, Reeve. Conch. Icon., t. 58, f. 391.   No. 2.

THAUMASTUS.

(See also figs. 9, 10.)

Fig. 14. T. CALIFORNICUS, Reeve. Conch. Icon., No. 378, Dec., 1848.
    W. G. Binney, l. c. iv., p. 24, t. 79, f. 15, 1859.   No. 3.

" 15. T. PATRIARCHA, W. G. Binney. Proc. Acad. Nat. Sci., p. 116, 1858.
    W. G. Binney, Terr. Moll., iv., p. 130, t. 80, f. 13, 1859.   No. 4.

" 16. T. ALTERNATUS, Say. New Harmony Disseminator, Dec. 28, 1831.
    W. G. Binney, l. c. iv., p. 126, t. 80, f. 1, 3, 1859.
    Bul. dealbatus, Say, (Part) Binney, l. c. ii., p. 276, t. 51, f. 2, t. 51 a, (excl. other figures,) 1851.
    Bul. lactarius, Menke.   No. 5.

ORTHALICIDÆ.

SYNONYMY AND REFERENCE TO PLATE 14.

THAUMASTUS.

(See also pl. 13.)

Figs. 1, 2, 4, 5. T. SCHIEDEANUS, Pfeiffer. Symb. ad Hist. Hel., i., p. 43, 1841.
    W. G. Binney, l. c. iv., p. 129, t. 80, f. 8, 15, 1859.
    Bul. Binneyanus, Pfr. W. G. Binney, l. c. iv., p. 128, 1859.
    Bul. dealbatus, (Part) Binney, l. c. ii., p. 276, t. 51, b., (excl. other figures,) 1851.   No. 6.

[xxxv.]

ACICULA.

Fig. 13. A. ACICULA, Müller. Hist. Verm. ii. p. 150.
Reeve, Conch. Icon. t. 20, f. 111.
Forbes and Hanley, Brit. Moll. iv. p. 130, t. 128, f. 4. No. 1.

ZUA.

" 14. Z. SUBCYLINDRACEA, Chemnitz. Conch. Cab. ix. p. 167, t. 135, f. 1235.
Achatina lubrica, Müller. Binney, Terr. Moll. ii. p. 283, t. 52, f. 4 (1851).
Bulimus lubricoides, Stimpson. Shells of New England, p. 54. No. 1.

RUMINA.

" 15. R. DECOLLATA, Linnæus. Syst Nat. p. 1247.
Binney, Terr. Moll. ii. p. 280, t. 50, f. 1 (1851).
Bulimus mutilatus, Say. Jour. Acad. Nat. Sci. ii. p. 373. No. 1.

MELANIELLA.

" 16. M. GRACILLIMA, Pfeiffer. Wiegman's Archiv. für Naturgesch. i. p. 352 (1839).
Binney, Terr. Moll. ii. p. 293, t. 53, f. 3 (1851).
Achatina striato-costata, D'Orbigny. Moll. Cuba, i. p. 176, t. 11, f. 21. No. 1.

OPEAS

" 17. O. SUBULA, Pfeiffer. Wiegman's Archiv. für Naturgesch. i. p. 352 (1839).
Binney, l. c. p. 285, t. 53, f. 4 (1851).
Bulimus octonoides, D'Orbigny. Moll. Cuba, i. p. 177, t. 11, f. 23, 24. No. 1.

MACROCERAMUS.

" 20. M. KIENERI, Pfeiffer. Proc. Zool. Soc. p. 40 (1846).
Reeve, Conch. Icon. Bulimus, t. 66, No. 463.
Pupa pontifica, Gould. Bost. Proc. iii. p. 40 (1848).
Cylindrella pontifica, Gould. Binney, l. c. ii. p. 306, t. 69, f. 1 (1851). No. 1.

Fig. 18, 19. M. Gossei, Pfeiffer. Proc. Zool. Soc. p. 137 (1845).
 Reeve, Conch. Icon. Bulimus, t. 66, f. 462.   No. 2.

PUPADÆ.

Synonymy and Reference to Plate 15.

ZOOGENITES.

" 1. Z. harpa, Say. Exped. St. Peters, ii. p. 256 (1824).
 Bulimus harpa, Binney, l. c. ii. p. 290, t. 52, f. 3 (1851).
 Zoogenites harpa, Morse. Shells of Maine, p. 32 (1864).
 Pupa costulata, Mighels, Bost. Proc. i. p. 187.   No. 1.

PUPILLA.

" 2. P. badia, Adams. Bost. Jour. iii. p. 331, t. 13, f. 18.
 Binney, l. c. ii. p. 323, t. 70, f. 3 (1851).
 = *Pupa muscorum*, Linn. *teste* Pfeiffer. Monog. Helic. Viv.   No. 1.

" 3. P. Hoppii, Möller. Index Moll. Grœnl. p. 4 (1842).
 W. G. Binney, l. c. iv. p. 147, t. 78, f. 20, 1859.   No. 2.

" 4. P. Blandi, Morse. Ann. N. Y. Lyc. Nat. Hist. viii. p. 211 (1865).   No. 3.

" 5. P. variolosa, Gould. Proc. Bost. Soc. iii. p. 40 (1848).
 Binney, l. c. ii. p. 331. t. 72, f. 2 (1851).   No. 4.

" 6. P. pentodon, Say. Jour. Acad. Nat. Sci. ii. p 376.
 Binney, l. c. ii. p. 328, t. 72, f. 1 (1851).
 Pupa curvidens, Gould. Invert. Mass. p. 189.
 Pupa Tappaniana, Adams. Moll. Vermont.   No. 5

Fig. 7. P. DECORA, Gould. Bost. Proc. ii. p. 263 (1847).
Binney, l. c. ii. p. 327, t. 71, f. 3 (1851). No. 6.
" 8. P. ROWELLI, Newcomb. Ann. Lyc. Nat. Hist. vii. p. 146.
Bland, Ann. Lyc. Nat. Hist. viii. p. 166 (1865). No. 7.
" 9. P. CALIFORNICA, Rowell. Ann. Lyc. Nat. Hist. vii. p. 287.
Bland, Ann. Lyc. Nat. Hist. viii. p. 166 (1865). No. 8.

LEUCOCHILA.

" 11. L. MARGINATA, Say.
Cyclostoma marginata, Say. Jour. Acad. Nat. Sci. ii. p. 72.
Bulimus marginatus (pars), W. G. Binney, l. c. iv. p. 136 (1859).
Bulimus fallax, Gould (pars). Binney, l. c. ii. p. 288 (1851).
Pupa albilabris, Adams. Moll. Vermont.
Bulimus marginatus, Say. Tryon, Am. Jour. Conch. i. p. 286 (1865). No. 1.

" 10. L. FALLAX, Say. Jour. Acad. Nat. Sci. v. (1825).
Bul. marginatus (pars), W. G. Binney, l. c. iv. p. 136 (1859).
Pupa fallax, Gould (pars). Binney, l. c. ii. p. 288 (1851).
Bulimus fallax, Say. Tryon, Am. Jour. Conch. i. p. 286 (1865). No. 2.

" 12. L. ARIZONENSIS, Gabb. Am. Jour. Conch. ii. p. 331, t. 21, f. 6 (1866). No. 3.

" —. L. HORDACEA, Gabb. Am. Jour. Conch. ii. p. 331, t. 21, f. 7 (1866). No. 4.

" 14. L. MODICA, Gould. Proc. Bost. Soc. iii. p. 40 (1848).
Binney, l. c. ii. p. 319, t. 52, f. 2 (1851). No. 5.

" 15. L. ARMIFERA, Say. Jour. Acad. Nat. Sci. ii. p. 162.
Binney, l. c. ii. p. 320, t. 70, f. 4 (1851). No. 6.

" 16. L. CONTRACTA, Say. Jour. Acad. Nat. Sci. ii. p. 374.
Binney, l. c. ii. p. 324, t. 70, f. 2 (1851).
Pupa deltostoma, Charpentier. Chemnitz, Conch. Cab. nov. edit. p. 181, t. 21, f. 17—19. No. 7.

Fig. 17. L. RUPICOLA, Say. Jour. Acad. Nat. Sci. ii. p. 163.
    Binney, l. c. ii. p. 341, t. 70, f. 1 (1851).
    Pupa procera, Gould. Bost. Jour. iii. p. 401, t. 3, f. 12.
    Pupa carinata, Gould. Cover of Bost. Jour. iv. Part 1.
    Pupa gibbosa, Chemnitz. Edit. 2, p. 123, t. 16, f. 13—16.
    Pupa minuta, Say. Pfeiffer, Monog. Hel. ii. p. 356 (1848).   No. 8.

" 18. L. CORTICARIA, Say (*Odostomia*). Nicholson's Encyc. iv. t. 4, f. 5.
    Binney, l. c. ii. p. 339, t. 72, f. 4 (1851).   No. 9.

" 26. L. PELLUCIDA, Pfeiffer. Symbolæ, i. p. 46.
    Küster, Chemnitz Conch. Cab. p. 89, t. 12, f. 24, 25.   No. 10.

STROPHIA.

" 19. S. INCANA, Binney. Terr. Moll. iii. t. 68 (1851).
    W. G. Binney, l. c. iv. p. 141 (1859).
    Pupa maritima (not of Pfeiffer). Gould in Terr. Moll. l. c. ii p. 316 (1851).
    Pupa detrita, Shuttleworth. Bern. Mittheil.   No. 1.

VERTIGO.

" 20. V. GOULDI, Binney. Bost. Proc. i. p. 105 (1843).
    Binney, Terr. Moll. ii. p. 332, t. 71, f. 2 (1851).   No. 3.

" 21. V. MILIUM, Gould. Bost. Jour. iii. p. 402, t. 3, f. 23.
    Binney, l. c. ii. p. 337, t. 71, f. 1 (1851).   No. 4.

" 22. V. OVATA, Say. Jour. Acad. Nat. Sci. ii. p. 375.
    Binney, l. c. ii. p. 334, t. 71, f. 4 (1851).
    modesta, Say.
    Pupa ovulum, Pfeiffer (olim). Symbolæ, i. p. 46.   No. 5.

" 23. V. SIMPLEX, Gould. Bost. Jour. iii. p. 403, t. 3, f. 21.
    Binney, l. c. ii. p. 343, t. 72, f. 3 (1851).   No. 6.

Fig. 24. V. CORPULENTA, Morse. Ann. N. Y. Lyc. viii.
p. 210 (1865). No. 2.
" 25. V. BOLLESIANA, Morse. Ann. N. Y. Lyc. viii.
p. 209 (1865). No. 1.
" 26. V. VENTRICOSA, Morse. Ann. N. Y. Lyc. viii.
p. 207 (1865). No. 7.

GONGYLOSTOMA.

" 27. G. POEYANA, D'Orbigny. Moll. Cuba, i. p. 185, t. 12, f. 24—26.
Cylindrella lactaria, Gould. Binney, l. c. iii. t. 69, f. 2.
C. lactaria of *text* in Binney, ii. p. 309 = *variegata*, Pfeiffer. No. 1.
" 28. G. JEJUNA, Gould. Bost. Proc. iii. p. 41 (1848). Binney, l. c. ii. p. 310, t. 69, f. 3 (1851. No. 2.
" 29. G. COAHUILENSIS, W. G. Binney. Am. Jour. Conch. i. p. 50. t. 7, f. 4, 5 (1865.) No. 3.

HOLOSPIRA.

—— H. ROEMERI, Pfeiffer. Monog. Hel. Viv. ii. p. 383. No. 1.
" 30. H. IRREGULARIS, Gabb. Am. Jour. Conch iii. p. 238, t. 16, f. 4 (1857).
W. G. Binney, lc. iv. p. 150 (1859). No. 5.
" 31. H. GOLDFUSSI, Menke. Zeit. für Malak. p. 2 (1847).
W. G. Binney, l. c iv. p. 151, t. 79, f. 33 (1859). No. 2.
" 32. H. REMONDI, Gabb. Am. Jour. Conch. i. (1865). No. 3.
" 33. H. NEWCOMBIANA, Gabb. Am. Jour. Conch. iii. p. 237, t. 16, f. 3 (1867). No. 6.
" 34. H. PFEIFFER, Menke. Zeitsch für Malakologie, p. 1 (1847). No. 4.

SNAILS.

Synonymy and Reference to Plate 16.

Fig. 1. Limax Columbianus, Gould. Binney, l. c. ii. p. 43, t. 66, f. 1 (1852).
Mollusca, Wilkes' Expl. Exped. p. 3, f. 1, a, b, c (1852). No. 5.

" 2. Limax maximus, Linn. Syst. Nat. ed. 10, i. 652 (1758). No. 2.

" 3. Limax flavus, Linnæus. DeKay, Moll. New York, p. 21, t. 1, f. 5 (1843).
Limax variegatus, Draparnaud. Binney, l. c. ii. p. 34, t. 65, f. 1 (1852). No. 1.

SNAILS.

Synonymy and Reference to Plate 17.

Fig. 1. Arion foliolatus, Gould. Binney, l. c. ii. p. 30, t. 66, f. 2 (1852).
Mollusca of Wilkes' Expl. Exped. p. 2, f. 2, a, b (1852). No. 2.

" 2—5. Veronicella Floridana, Binney. Terr. Moll. ii. p. 17, t. 67 (1852). No. 1.

" 6. Tebennophorus Carolinensis, Bosc. p. 80 (1830).
Binney, l. c. ii. p. 20, t. 63, f. 1, 2 (1852).
Limax togata, Gould. Invert. Mass. p. 3 (1841). No. 1.

Fig. 7, 8. TEBENNOPHORUS DORSALIS, Binney. Limacidæ, p. 14, Bost. Jour. iv. p. 174 (1842).
    Binney, Terr. Moll. ii. p. 24, t. 53, f. 3 (1852).   No. 2.

" 9, 10. ARION FUSCUS, Müller.
    A. hortensis, Ferussac. Binney, l. c. ii. p. 27, t. 64, f. 1, t. 65, f. 2 (1852).   No. 1.

" 11, 12, 13. LIMAX CAMPESTRIS, Binney. Limacidæ, p. 9 (1842).
    Boston Journal, iv. p. 169 (1842).
    Binney, Terr. Moll. ii. p. 41, t. 64, f, 3 (1852).   No. 4.

" 14, 15, 16. LIMAX AGRESTIS, Müller. Hist. Verm. Part 2, p. 8.
    Binney, l. c. ii. p. 36, t. 64, f. 2 (1852).   No. 3.

OPERCULATA.

SYNONYMY AND REFERENCE TO PLATE 18.

Figs 1, 2. ALEXIA MYOSOTIS, Draparnaud.
    W. G. Binney, l. c. iv., p. 172, t. 75 f. 33, t. 79, f. 16, 1859.
    Auricula denticulata, Gould, Invert. Mass, p. 199, t. 129, 1841.
    Melampus borealis, Conrad, Am. Jour. Science. xxiii., p. 345, 1833.   No. 1.

" 3. CARYCHIUM EXIGUUM, Say. Jour. Acad. Nat. Sciences. ii., p. 375, 1822.
    Binney, Terr. Moll. ii., p. 286, t. 53, f. 1, 1852.
    C. exile, H. C. Lea, Am. Jour. Science, xlii. p. 109, t. 1, f. 5, 1841.
    C. existelium, Bourguignat, Mag. Zool. p. 220, 1857.
    C. euphæum, Bourguignat, Mag. Zool. p. 221, 1857.   No. 1.

MELAMPUS, Montfort.

" 4. M. OLIVACEUS, Carpenter. Cat. Reigen Coll. p. 178, 1857.   No. 1.

Fig. 5. M. BIDENTATUS, Say. Jour. Acad. Nat. Sc., ii., p. 245, July, 1822.
    W. G. Binney, l. c., iv., p. 156, t. 75, f. 23, 1859.
    M. corneus, Stimpson, Shells of New England, p. 51, 1851.
    Auricula cornea, Deshayes, Encyc. Meth., ii., p. 90, 1830.
    Auricula biplicata, Deshayes, Encyc. Meth., ii., p. 91, 1830.
    Auricula jaumei, Mittre, Rev. Zool. p. 66, 1841.   No. 2.

" 6. M. FLAVUS, Gmelin. Syst. Nat., p. 3436, 1788.
    W. G. Binney, l. c., iv., p. 166, 1859.
    Bulimus monile, Bruguiere, Encyc. Meth. i., p. 338, 1792.   No. 3.

" 7, 8. M. COFFEA, Linn. Syst. Nat. xii., edit. p. 1187.
    W. G. Binney, l. c., iv., p. 162, t. 75, f. 21, 25, 1859.
    Voluta minuta, (part) Gmelin, Syst. Nat., p. 3436, 1788.
    Ellobium Barbadense, Bolten, Museum.
    Bulimus coniformis, Bruguiere, Encyc. Meth. i., p. 339, 1792.
    Auricula ovula, D'Orbigny, Moll. Cuba, i., p. 187, t. 13, f. 4—7, 1853.   No. 4.

TRALIA, Gray.

" 9. T. PUSILLA, Gmelin. Syst. Nat., p. 3436, 1788.
    W. G. Binney, l. c., iv., p. 168, t. 75, f. 29, 1859.
    Voluta triplicata, Donovan, Brit. Shells, v., t. 138, 1808.
    Auricula nitens, Lamarck, Anim. Sans Vert, vi., p. 141.   No. 1.

" 10. T. CINGULATA, Pfeiffer. Wiegmann's Archiv. für natür, i., p. 251, 1840.
    W. G. Binney, l. c., iv., p. 161, t. 75, f. 12, 13, 1859.

[xliii.]

  Auricula oliva, D'Orbigny, Moll. Cuba, i., p. 189, t. 12, f. 8—10.   No. 2.

Fig. 11. T. Floridana, Shuttleworth. Pfeiffer. Malak. Blätt, 1854.
   W. G. Binney, l. c., iv., p. 165, t. 75, f. 30, 1859.   No. 3.

LEUCONIA, Gray.

" 12. L. Sayi, Küster. Conchylien Cabinet, Auricula, p. 42, t. 6, f. 14, 15, 1844.
   W. G. Binney, l. c., iv., p. 177, t. 75, f. 34, 1859.   No. 1.

PEDIPES, Adanson.

" 38. P. lirata, W. G. Binney, Proc. Acad. N. S.   No. 1.

BLAUNERIA, Shuttleworth.

" 13. B. pellucida, Pfeiffer. Wiegmann's Archiv für Naturg., i., p. 252, 1840.
   Binney, Terr. Moll. ii., p. 294, t. 53, f. 2, 1852.
  Tornatellina Cubensis, Pfeiffer, Symbolæ, ii., p. 130, 1842.   No. 1.

CHONDROPOMA, Pfeiffer.

" 14—16. Chondropoma dentatum, Say. Jour. Acad Nat. Sciences, v, p. 123, 1825.
   Binney, l. c. ii, p. 348, t. 62, 1852.
   W. G. Binney, l. c. iv, p. 191, t. 75, f. 24, 1859.   No. 1.

HELICINA, Lamarck.

" 17—19, 26. H orbiculata, Say. Jour. Acad. Nat. Sci. i, p. 283, 1818.
   Binney, l. c. ii, p. 352, t. 72, 74, f. 3, 1852.
  H. tropica, Jan. Chemnitz. Edit. 2, p. 37, t. 4, f. 9, 10, 1846–9.
   W. G. Binney, l. c. iv, p. 194, t. 73, middle figure, lower line.
  H. ambeliana, Sowerby. Thes. Conch. t. 1, f. 19, 1842.   No. 1.

Figs. 20, 21. H. occulta, Say. Jour. Acad. Nat.
Sci. i, p. 182, 1818.
Binney, l. c. ii, p 356, t. 74, f. 1, 2, 1852.
H. rubella, Green. Doughty's Cabinet of Nat.
Hist. ii, p. 291, 1832. No. 2.

" 22, 23. H. hanleyana, Pfeiffer. Proc Zool.
Soc. p. 122, 1848.
W. G. Binney, l. c. iv, t. 75, f. 14, 16, 1859. No. 3.

" 24. H. chrysocheila, Binney. l. c. ii, p. 354,
t. 74, f. 4, 1852. No. 4.

" 25. H. subglobulosa, Poey. Memorias, i, p.
115, 120, t. 12, f. 17—21.
W. G. Binney, l. c. iv, p. 195, t. 75, f. 17,
1859. No. 5.

TRUNCATELLA.

" 27, 28. T. caribæensis, Sowerby. Reeve's
Conch. Syst. ii, t. 182, f. 7.
W. G. Binney, l. c. iv, p. 185, t. 75, f. 2, 4,
1859.
T. succinea, C. B. Adams. Bost. Proc. 12,
1845. No. 1.

" 29—31. T. subcylindrica, Pulteney. Dorset.
Cat. p. 49, 1799.
W. G. Binney, l. c. iv, p. 186, t. 75, f. 5, 6,
8, 1859.
T. Caribæensis, Pfeiffer. Zeitschr. für Malak.
p. 182, 1846 (ex parte.) No. 2.

" 32, 33. T. bilabiata, Pfeiffer. Wiegmann's
Archiv für Naturg. i, p. 253, 1840.
W. G. Binney, l. c. iv, p. 188, t. 75, f. 3, 7,
1859. No. 3.

" 34—36. T. pulchella, Pfeiffer. Wiegmann's
Archiv für Naturg. i, p. 356, 1839.
W. G. Binney, l. c. iv, p. 189, t. 64, f. 1, 9,
19, 1859. No. 4.

" 37. T. californica, Pfeiffer. Proc. Zool. Soc.
p. 111, 1857.
W. G. Binney, l. c. iv, p. 28, t. 79, f. 20,
22, 1859. No. 5.

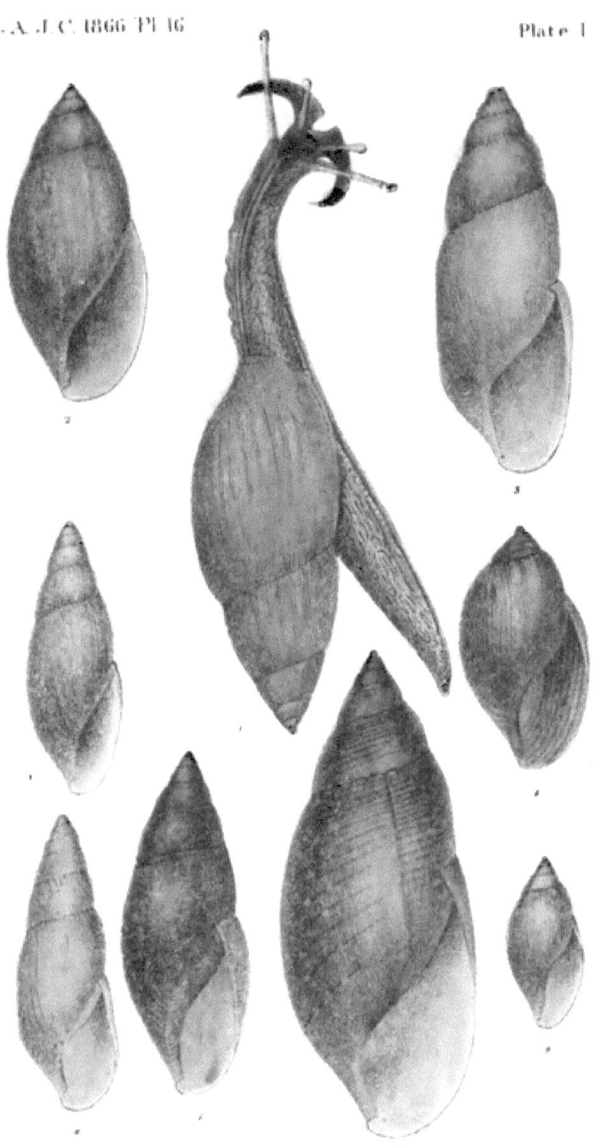

Drawn by E J Nolan   Bowen & Co lith. Phila da

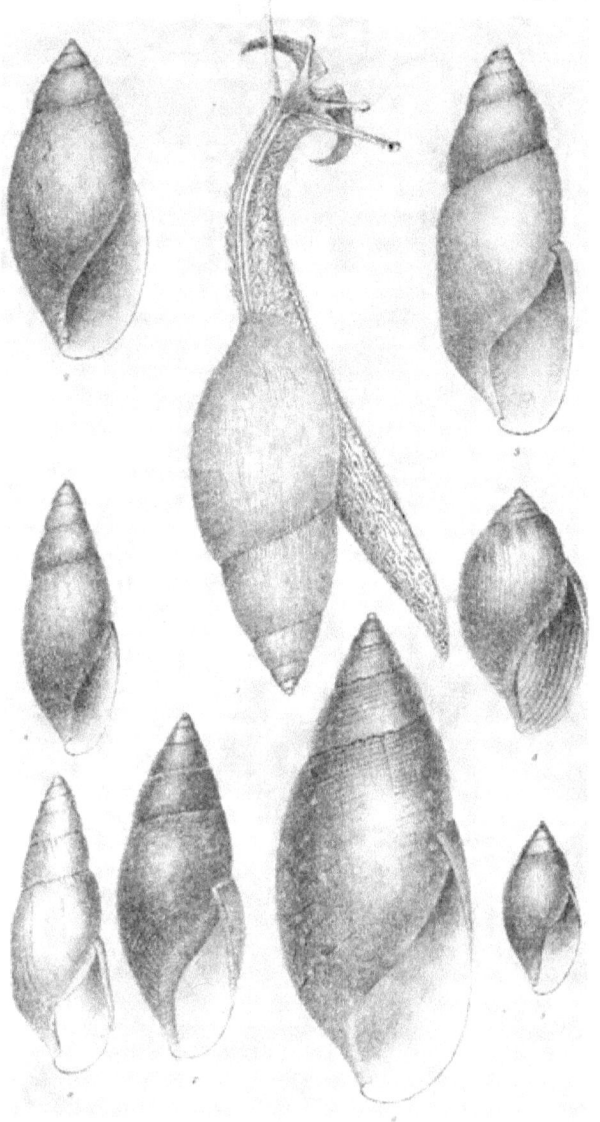

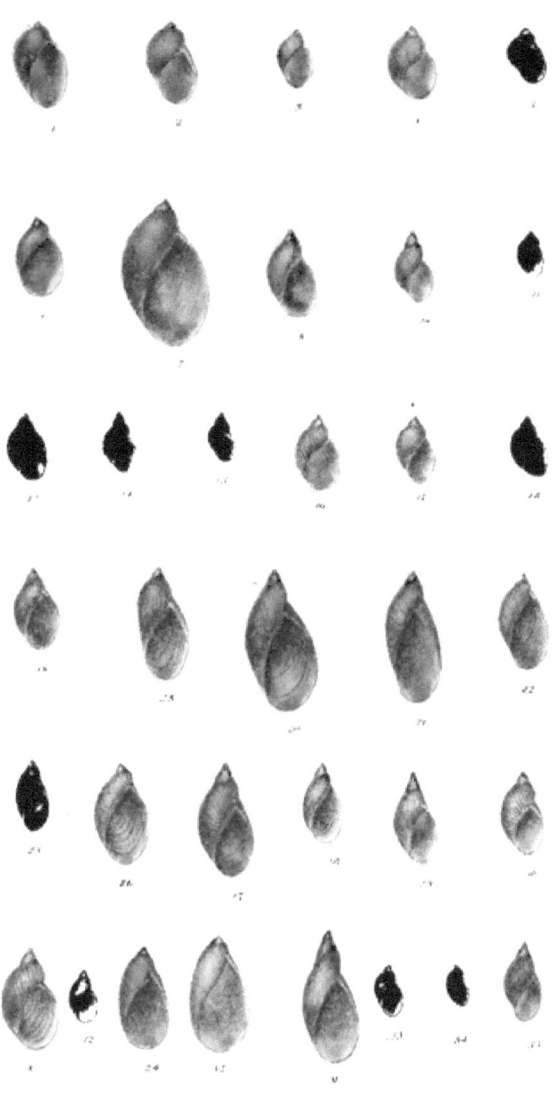

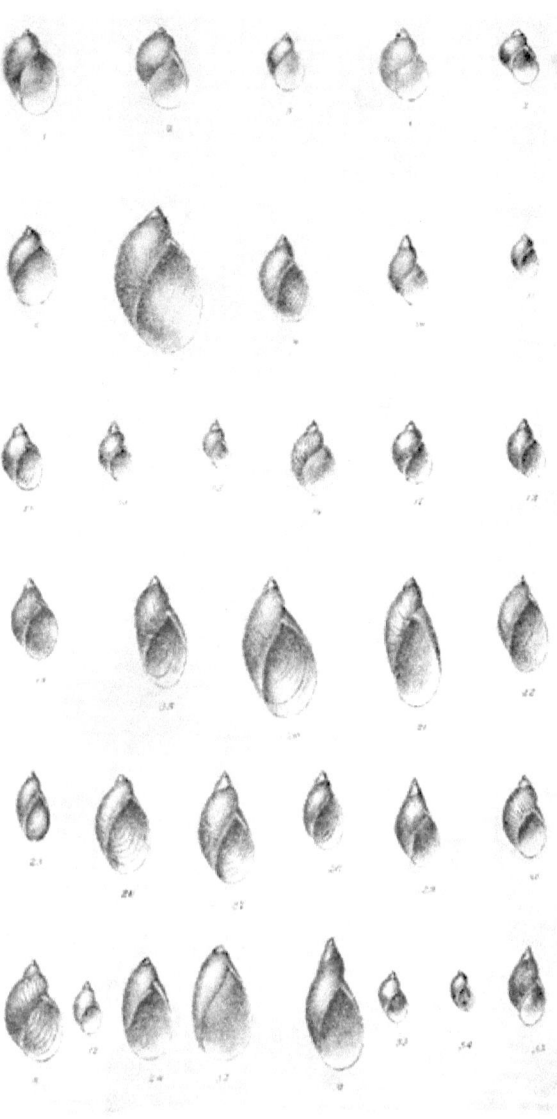

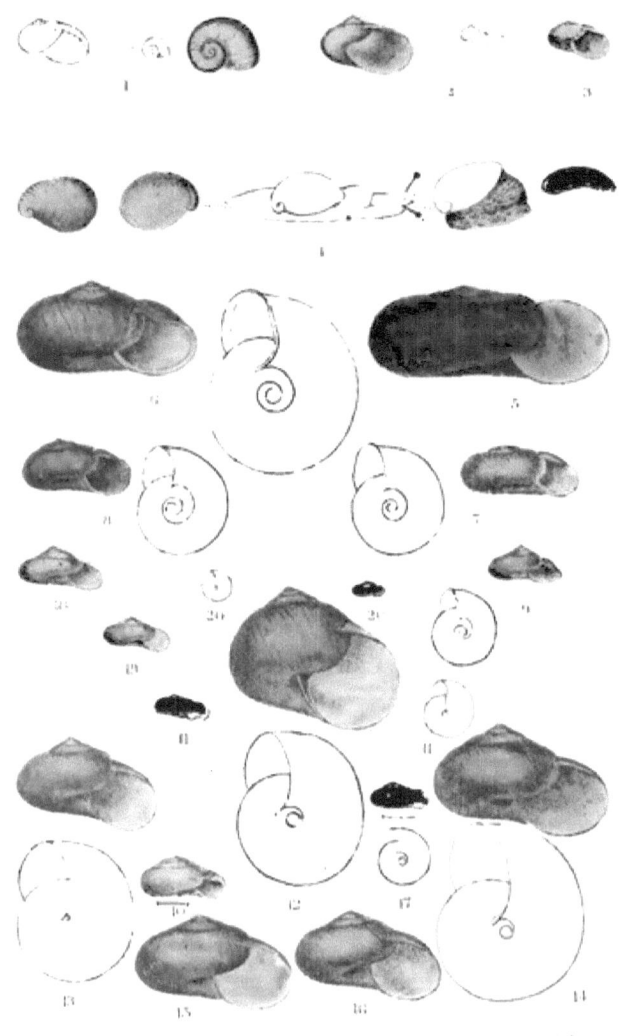

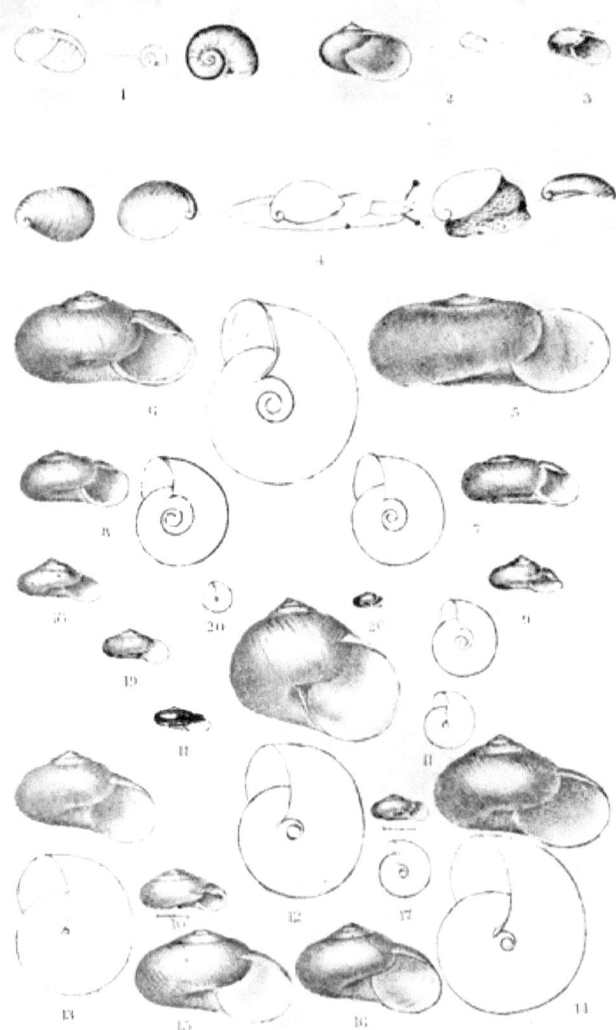

Plate IV

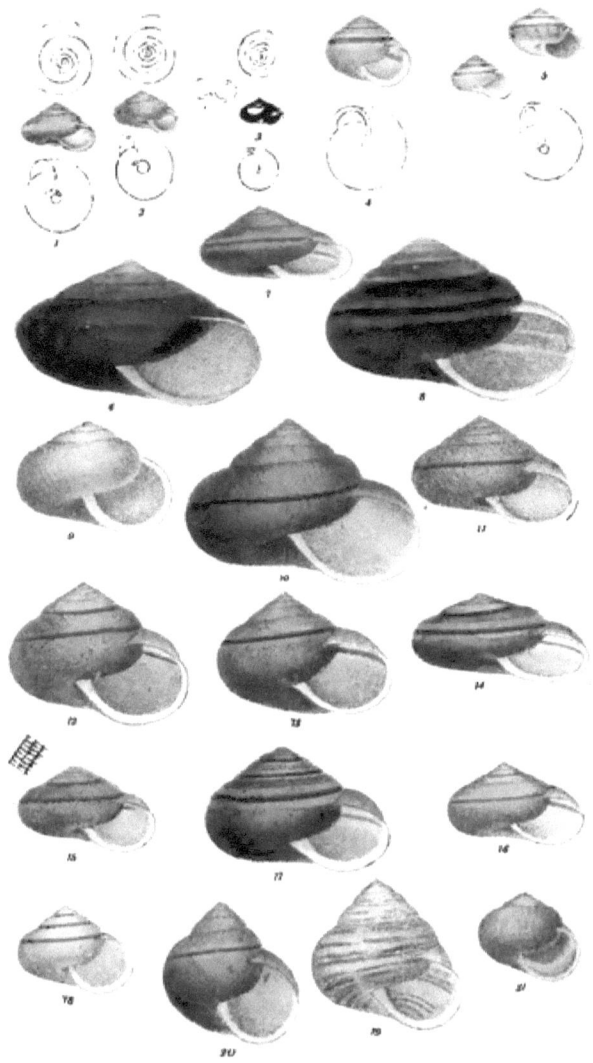

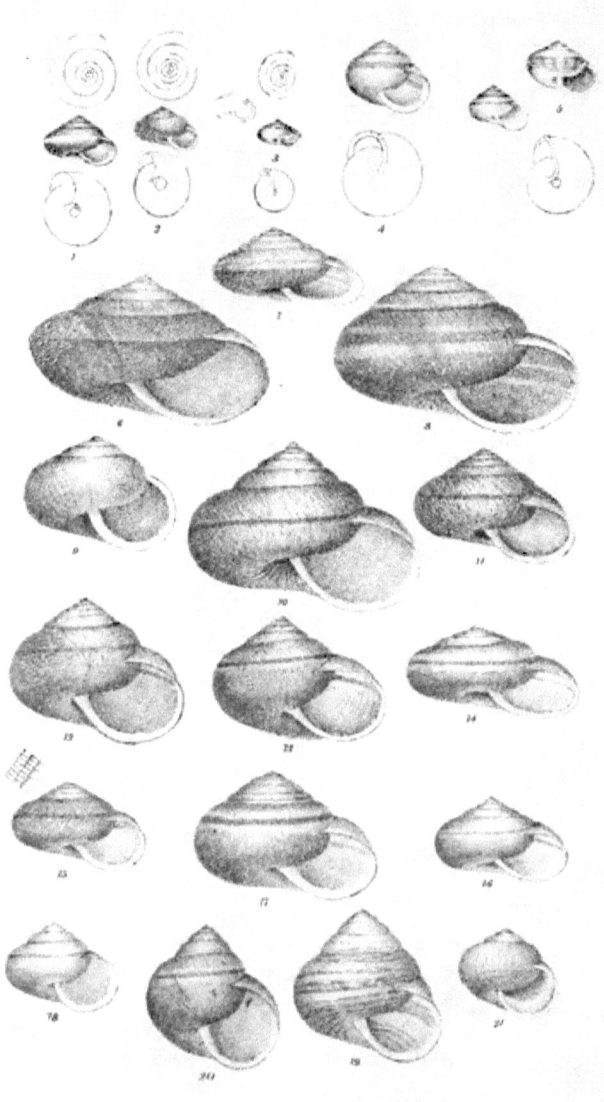

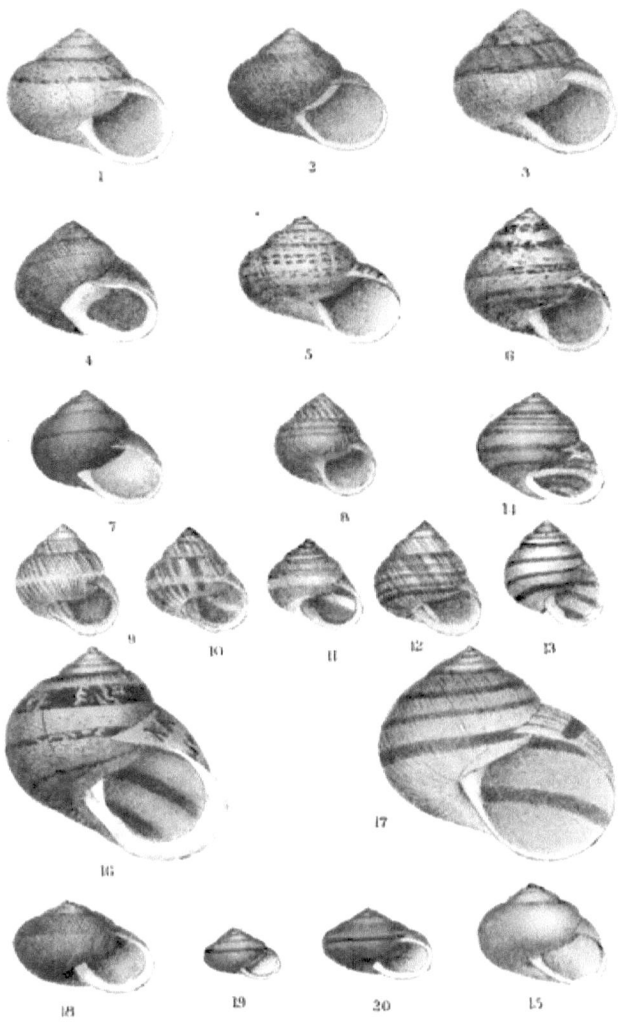

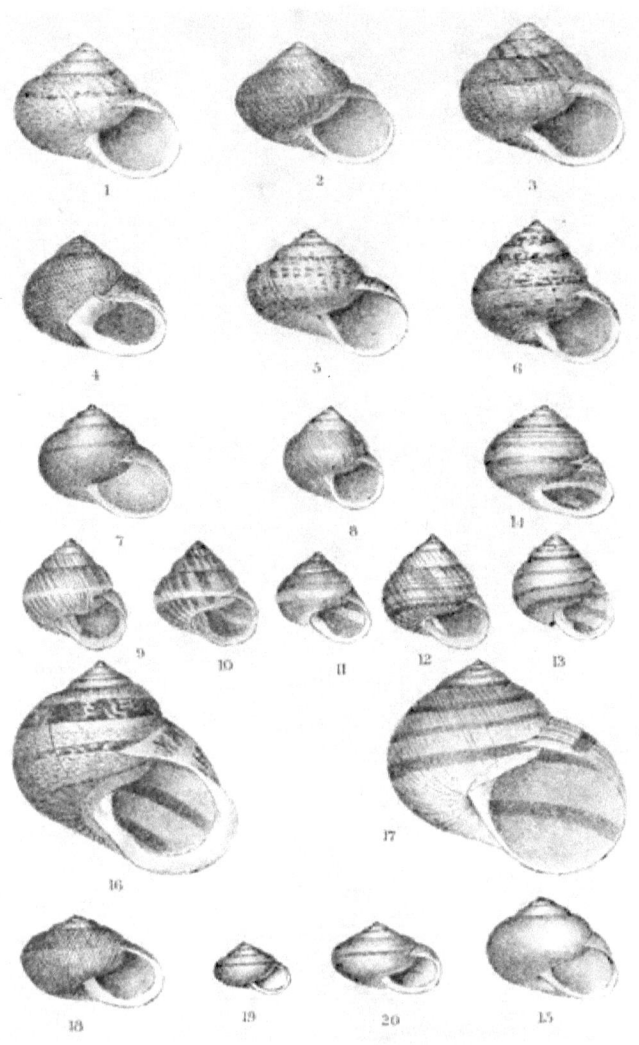

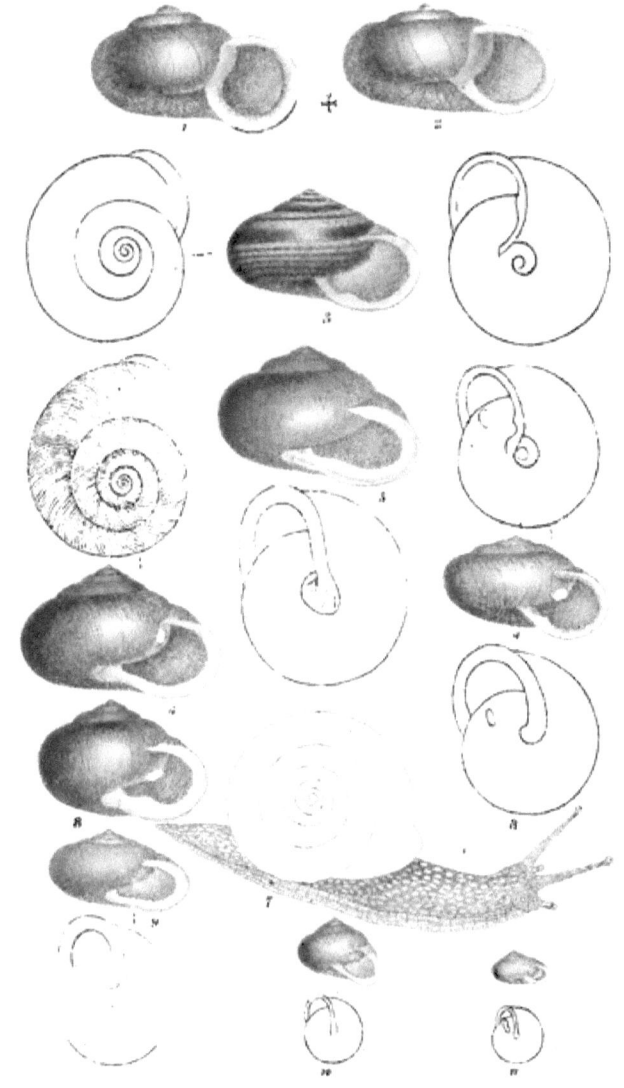

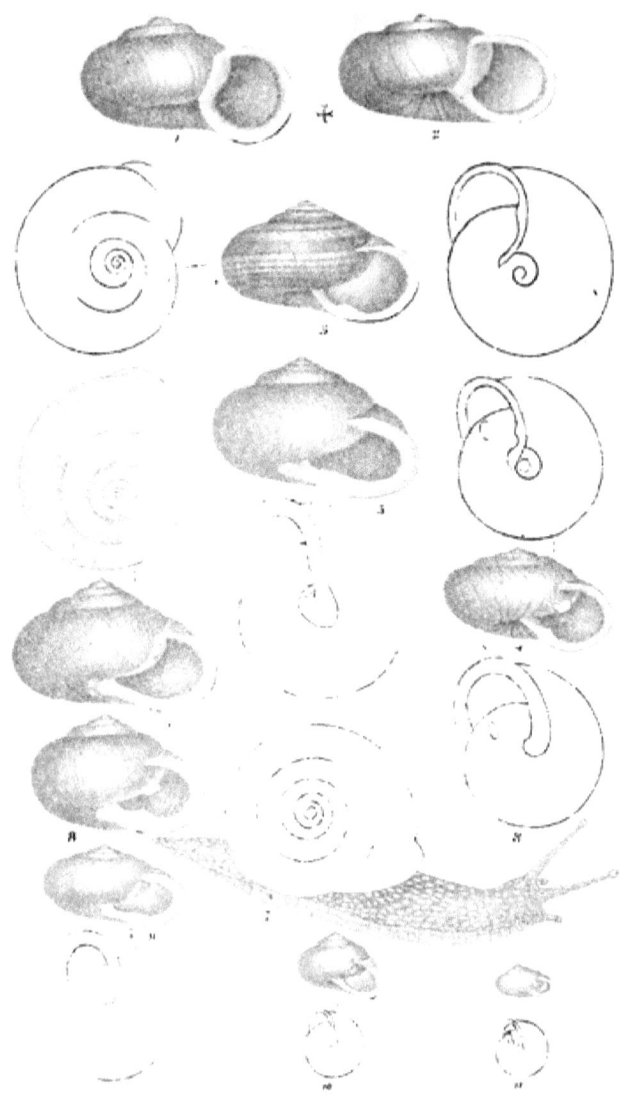

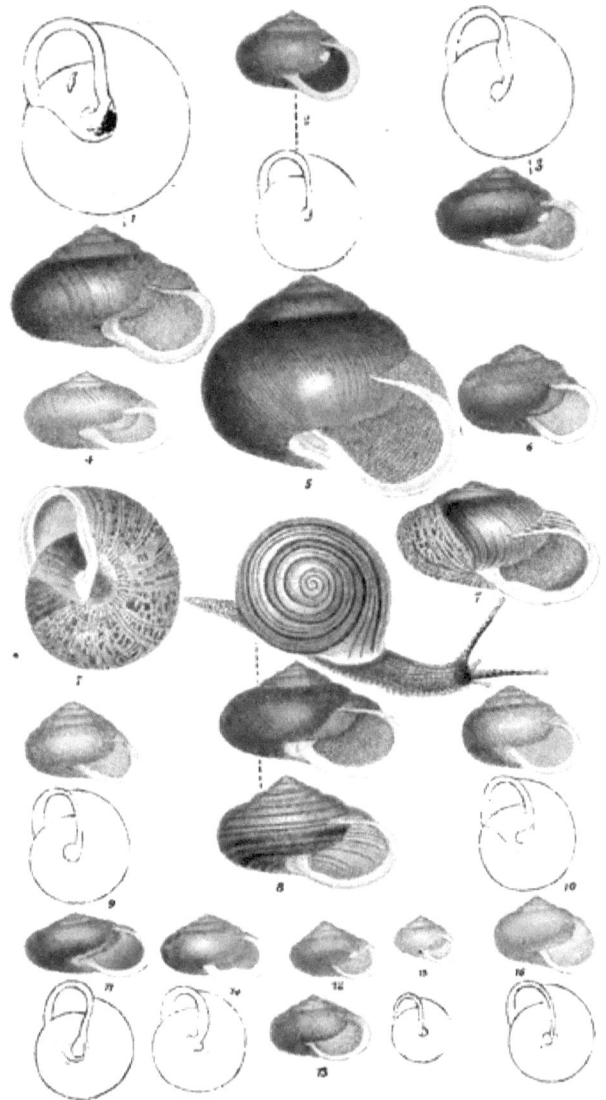

Plate 8.

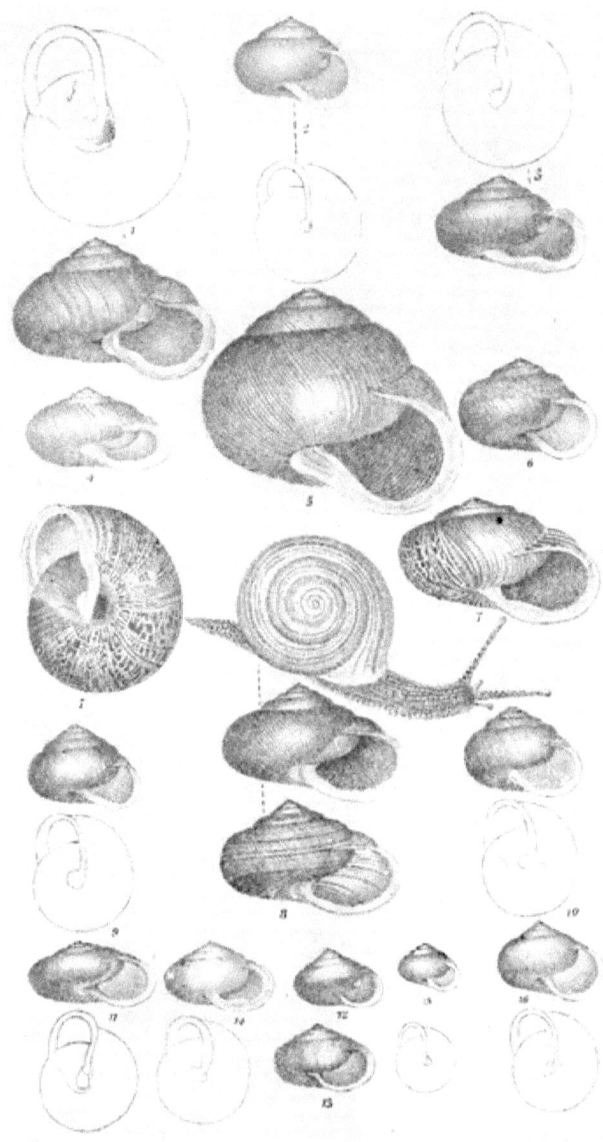

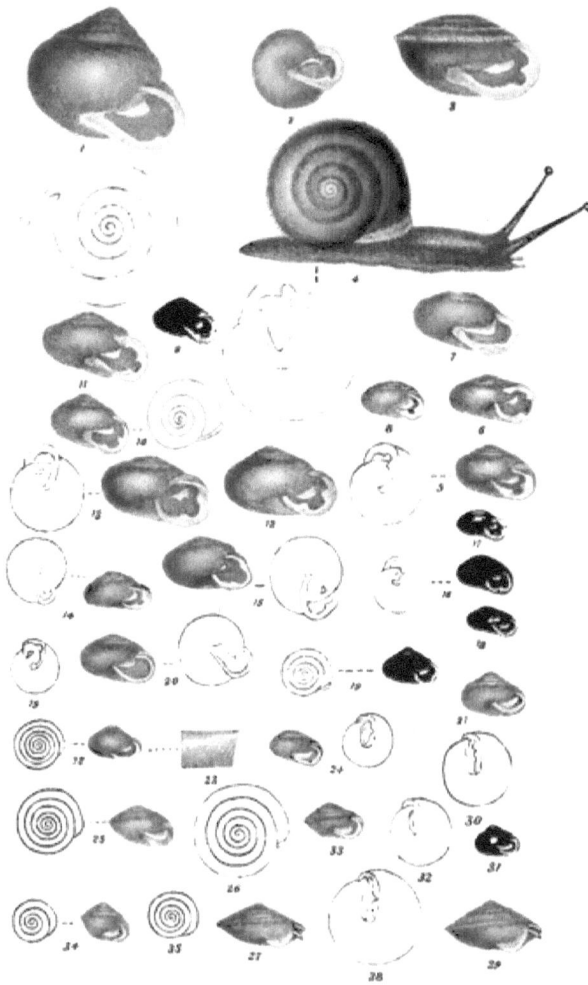

Plate 9

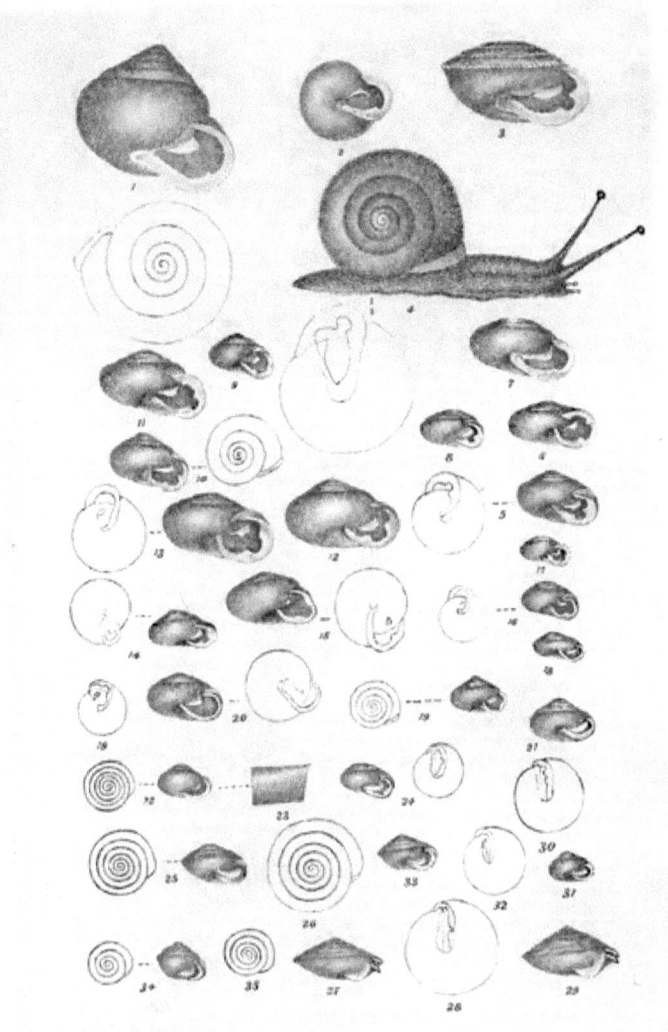

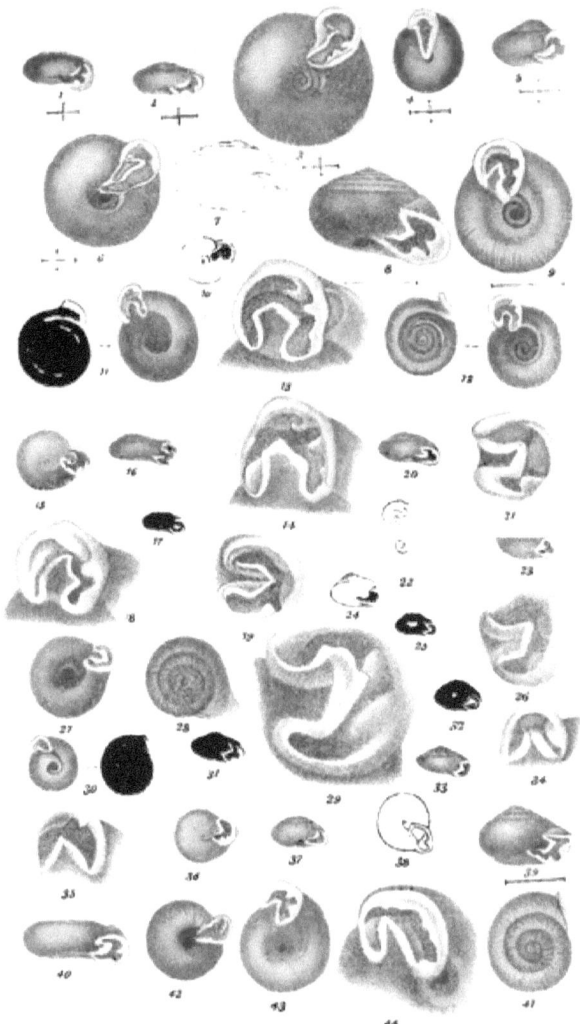

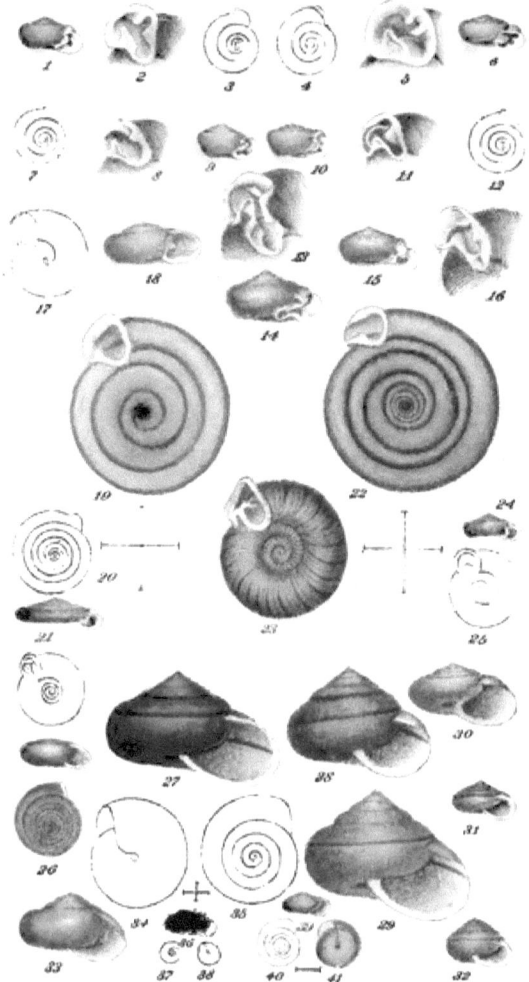

Drawn by E. J. Nolan, M.D. Bowen & Co. lith. Philada.

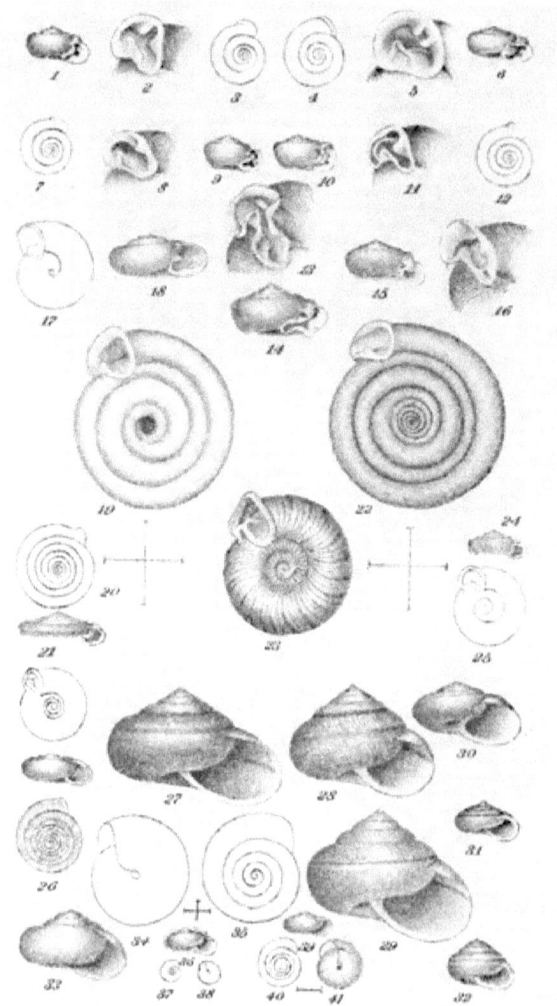

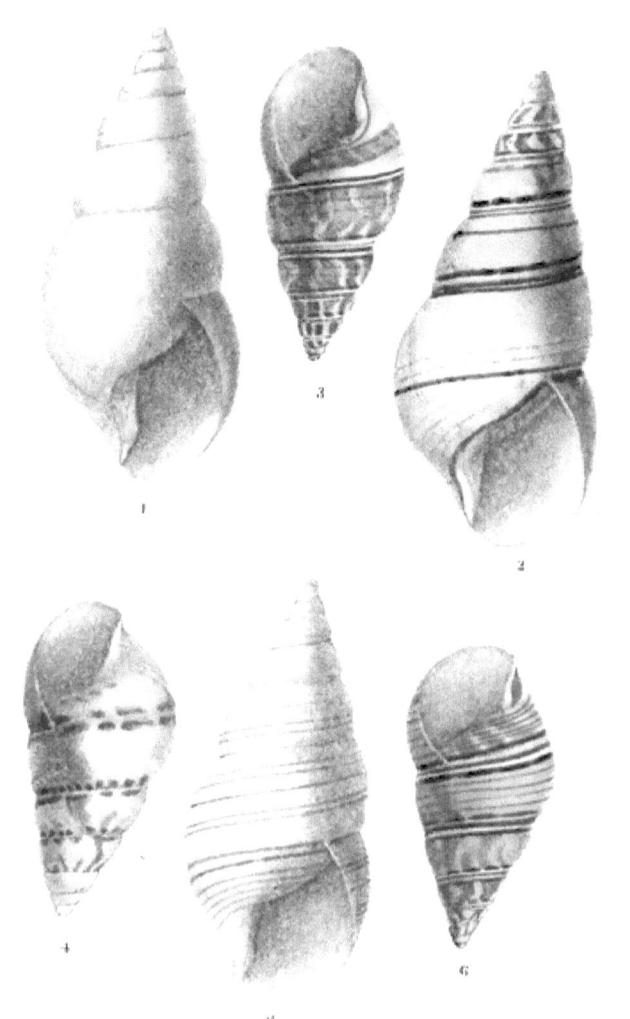

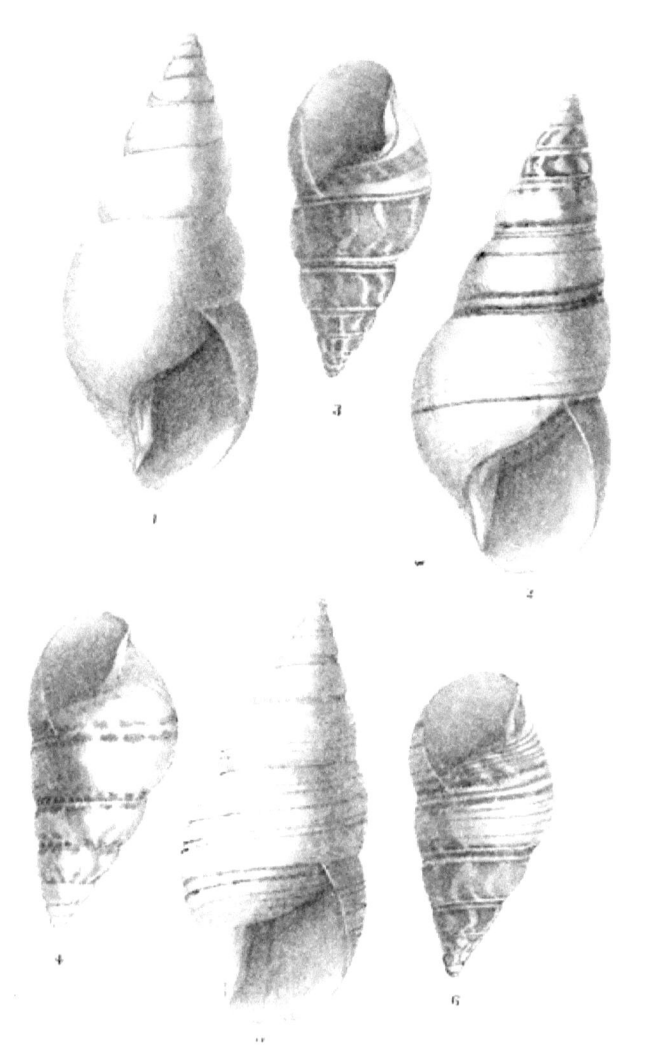

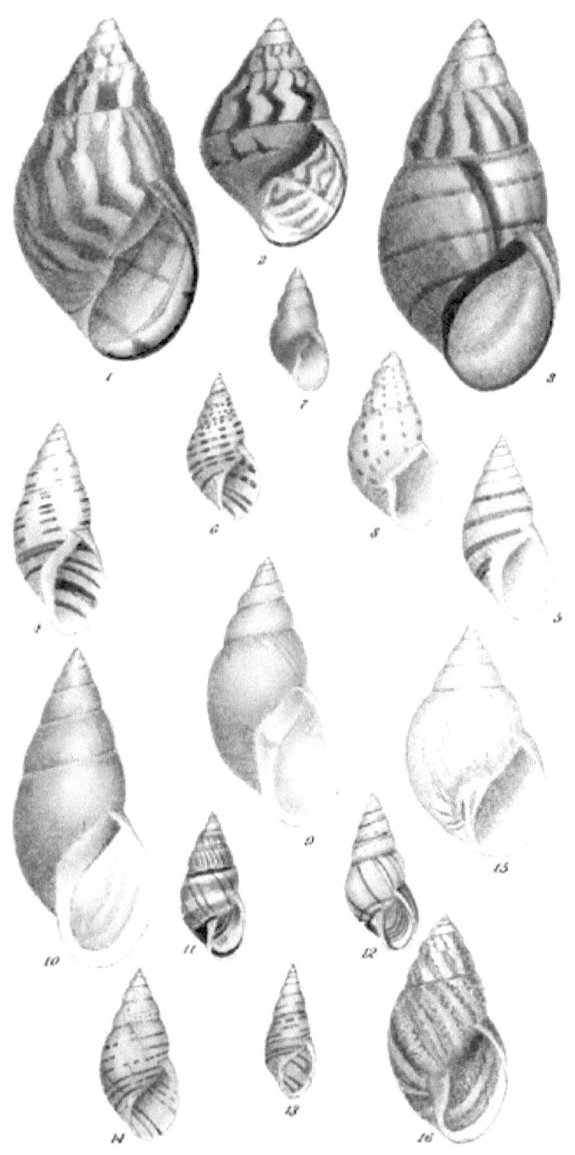

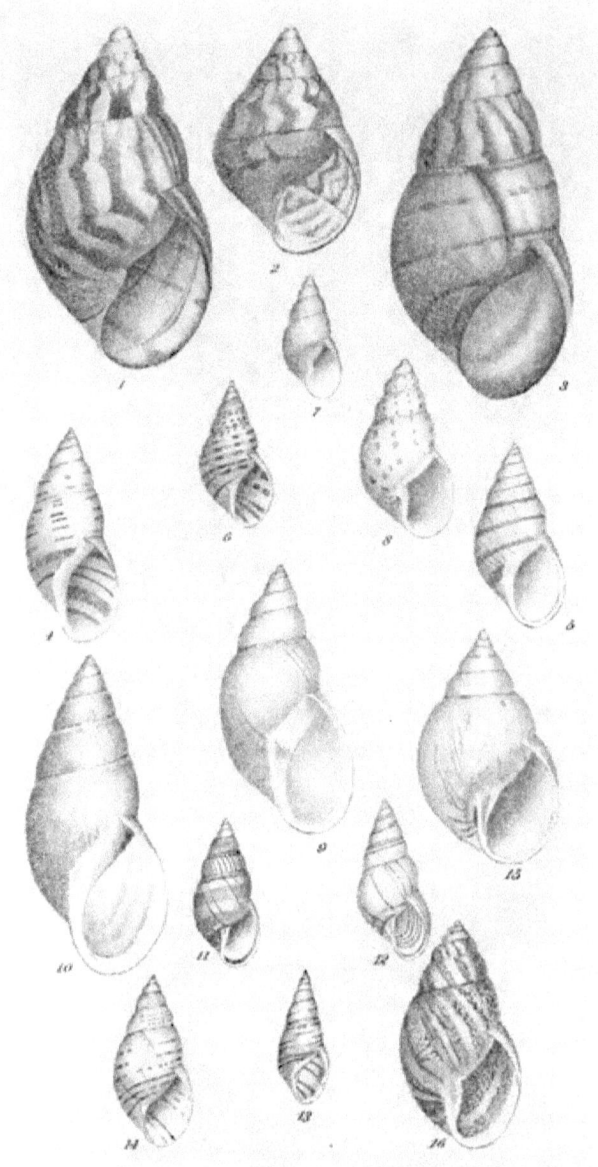

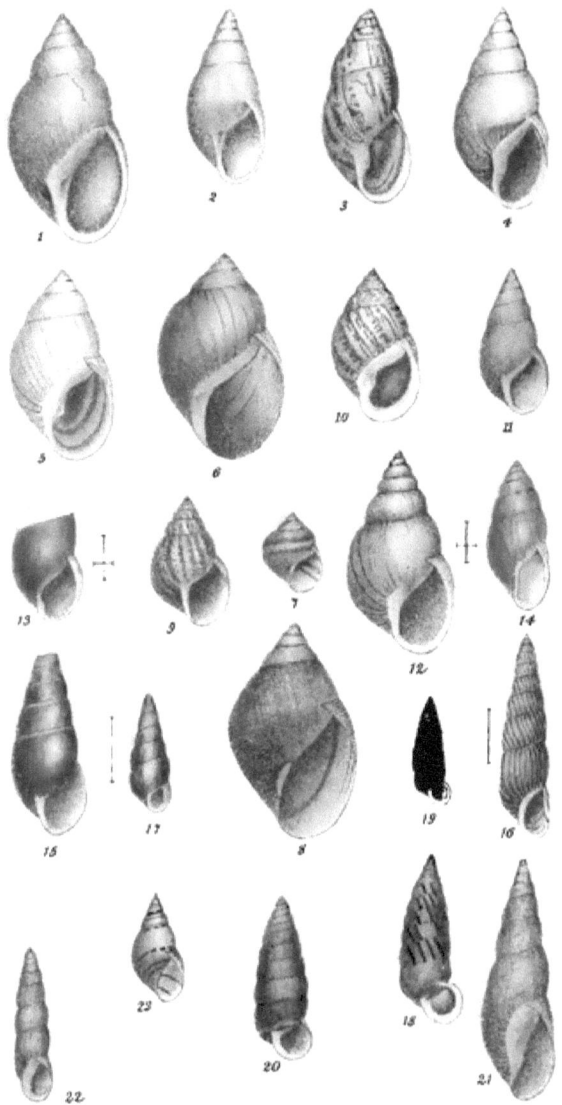

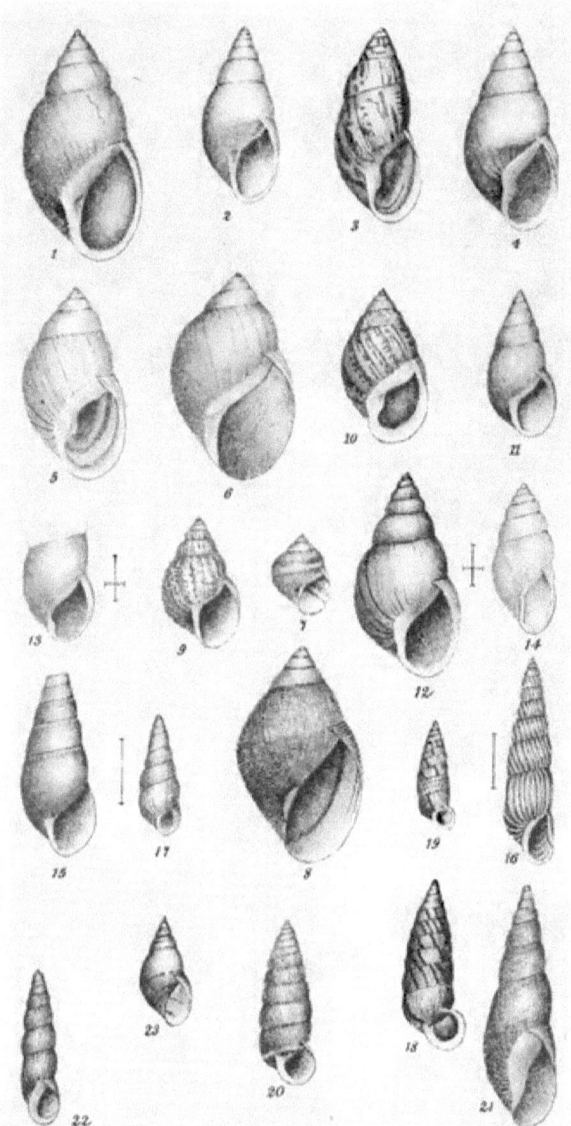

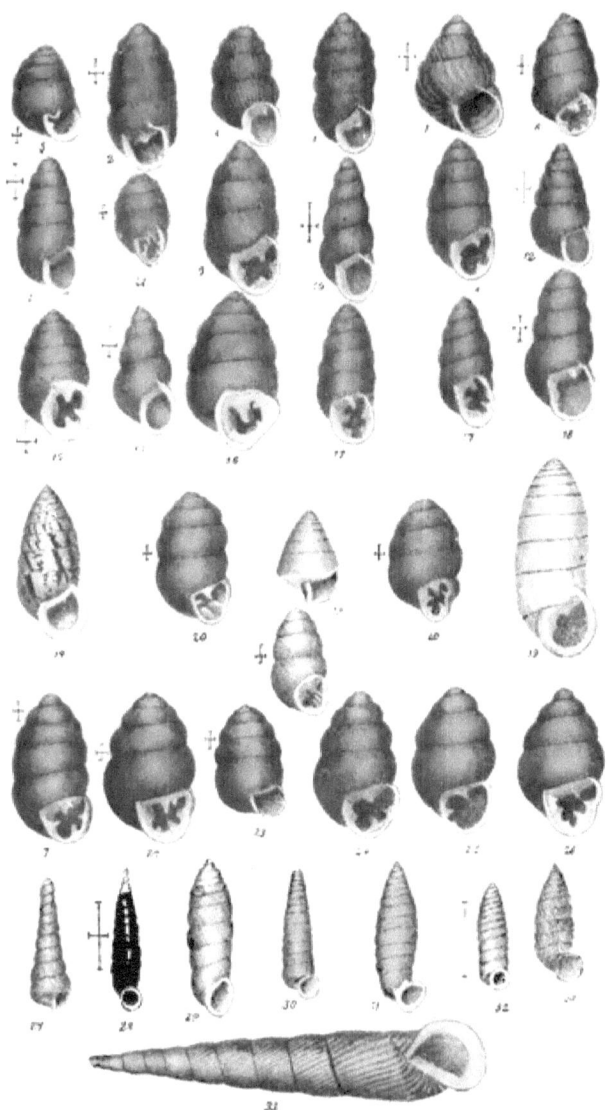

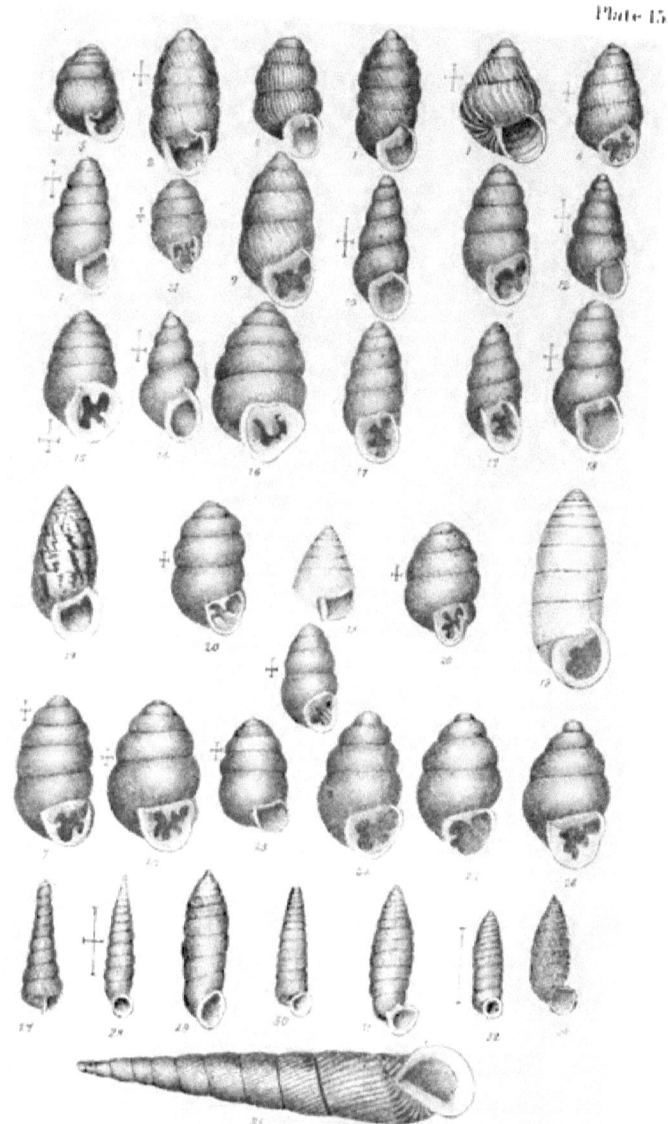

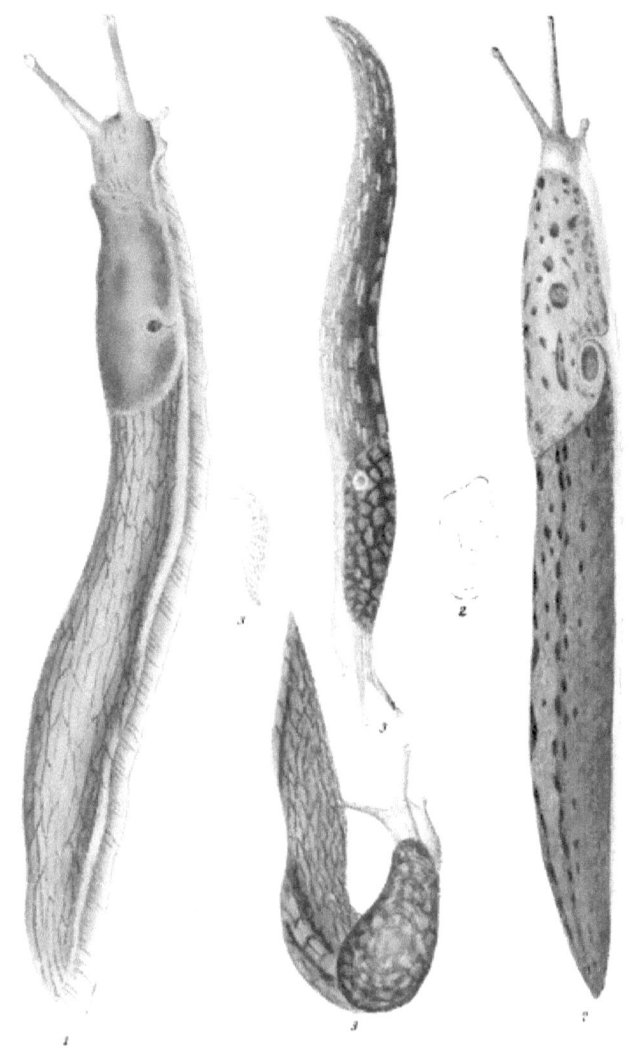

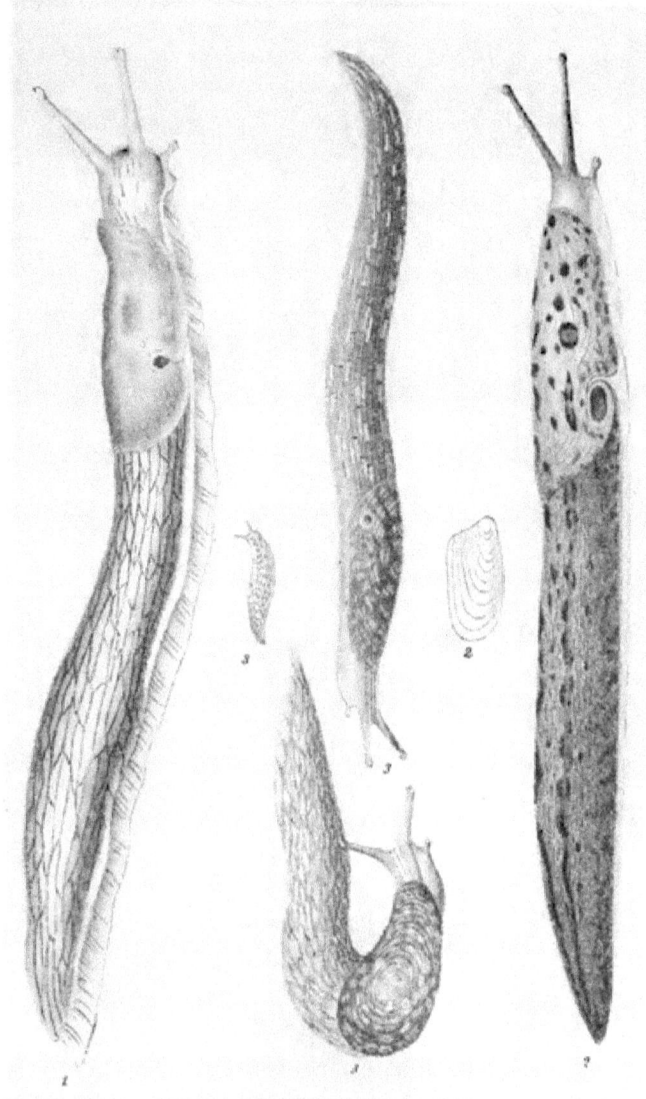

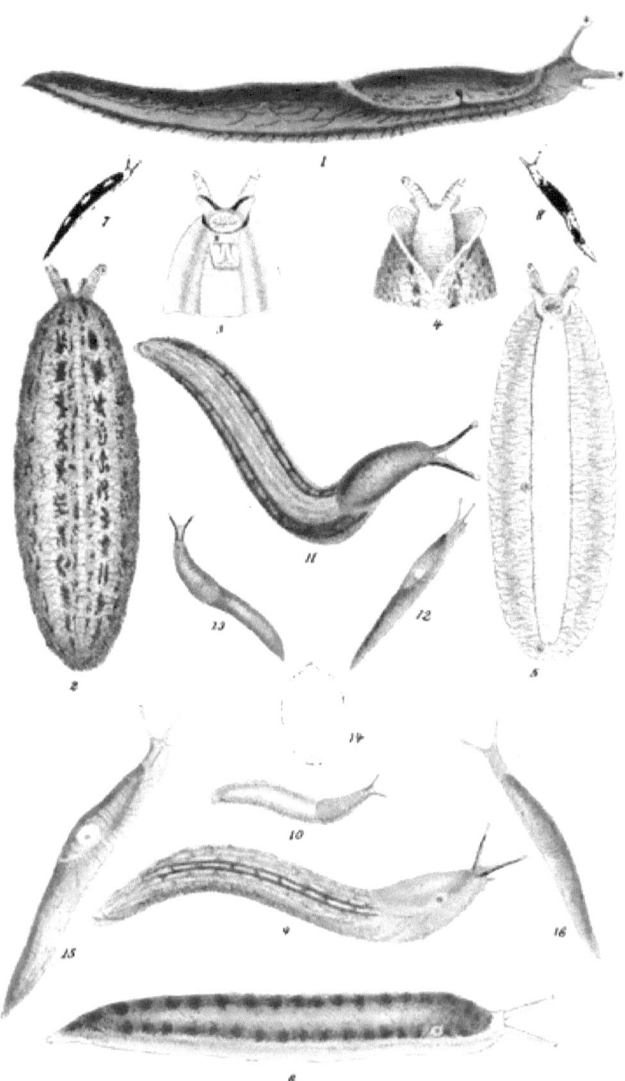

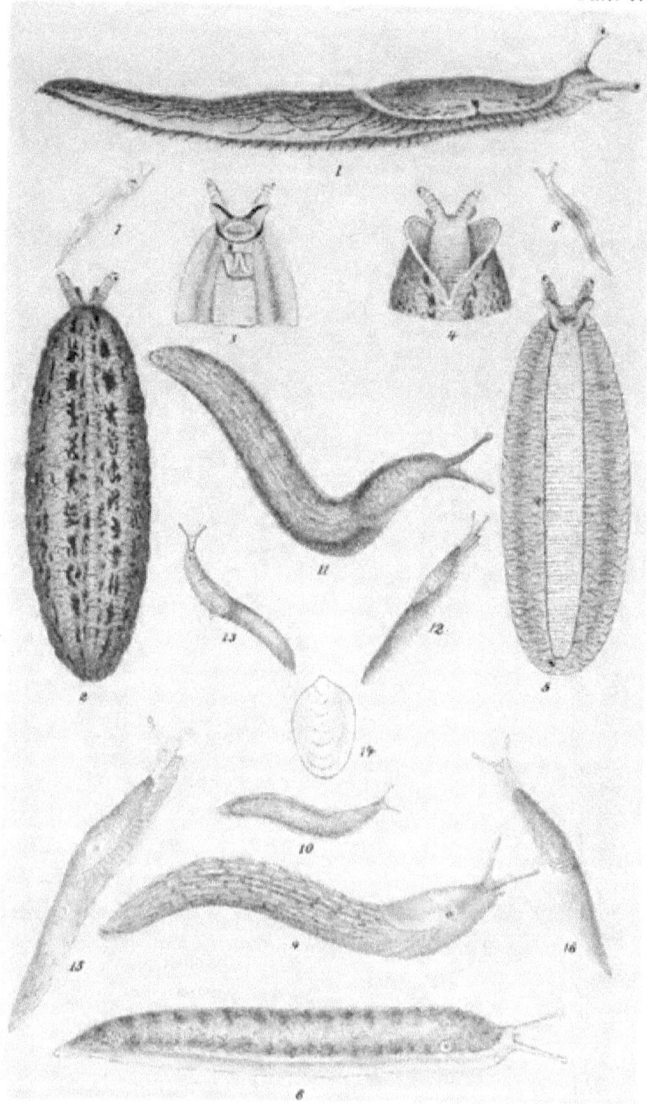

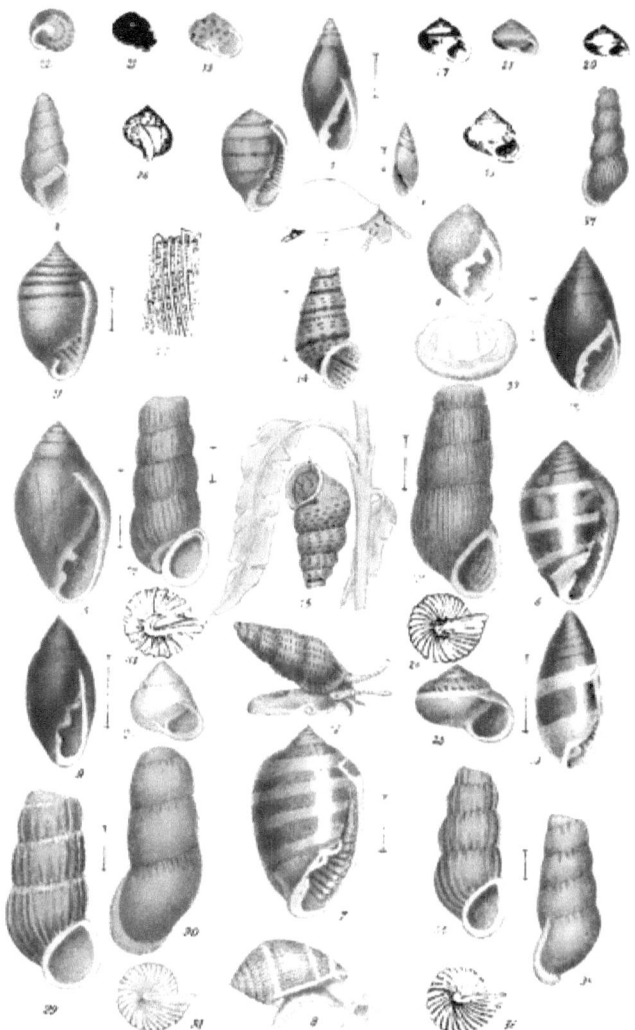

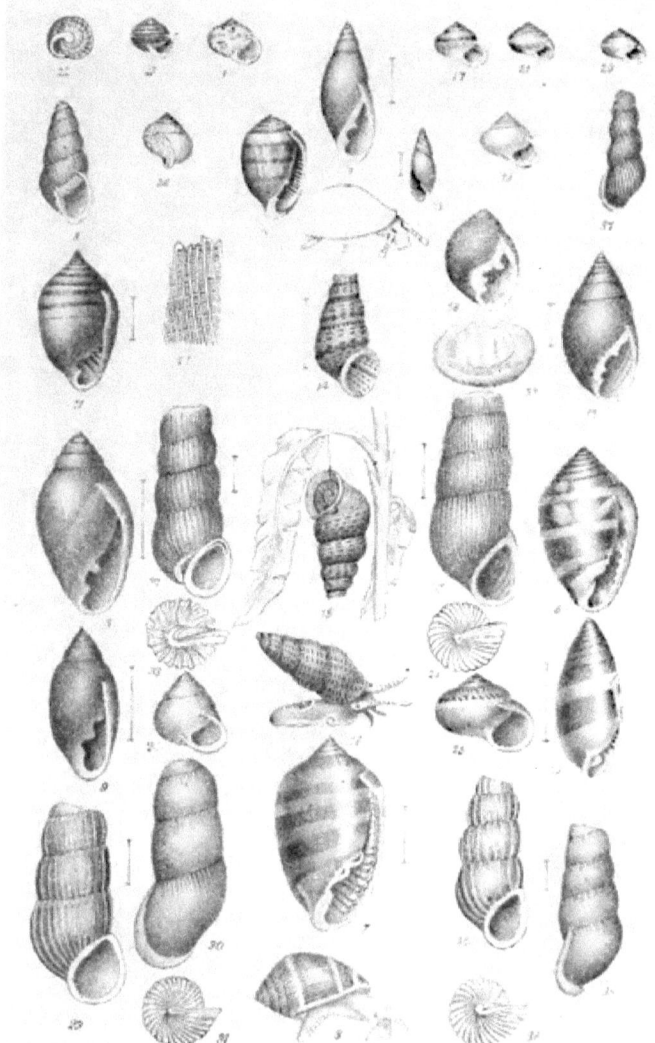